WATERSHED MANAGEMENT

FOR DRYLAND AGRICULTURE

WATERSHED MANAGEMENT
FOR DRYLAND AGRICULTURE

M. C. Oswal

2023

Associated Publishing Company
A Division of
Astral International Pvt. Ltd.
New Delhi – 110 002

First Published, 1999
Reprinted, 2023

ISBN: 978-93-5461-543-6 (Int. Edn.)

Published by : **Associated Publishing Company**
A Division of
Astral International Pvt. Ltd.
– ISO 9001:2015 Certified Company –
4736/23, Ansari Road, Darya Ganj
New Delhi-110 002
Ph. 011-43549197, 23278134
E-mail: info@astralint.com
Website: www.astralint.com

Preface

Dryland agriculture is the mainstay of over 76 per cent population in the developing countries of the world, located in tropics. Development of dryland agriculture is therefore essential to meet the growing demand of food, fodder, and fuel, narrowing down imbalance between irrigated and unirrigated areas, creating rural employment and maintaining favourable ecological balance.

The low yields (< 0.5 t ha^{-1}) of dryland areas is due to frequent failure of crops on account of uncertain rainfall and lack of scientific farming. Despite this, dry farming has been practised without essential inputs since time immemorial for subsistance. Consequently, the land has been degraded. This calls for taking effective measures for land and rain-water conservation and their efficient use for improving and sustaining dryland agriculture. It could be possible by adopting an integrated approach towards the management and development of natural resources including soil, water, crop, range, forest and livestock on the basis of watershed approach after involving the people of the area.

The importance of watershed development of dryland areas has been well recognized now and a few universities in developing countries have recently included this as a subject for regular teaching at graduate and undergraduate levels.

A few books on watershed are available in the market. These books are largely based on case studies and do not cover a complete spectrum of watershed management. This book is an attempt to fill this gap. The book presents twelve chapters. The first chapter covers introduction and development of watershed concept and its importance in dryland agriculture while second chapter describes the two basic natural resources—soil and climate and crop planning for dryland areas. The third chapter is exclusively devoted for watersheds objectives, planning and execution. Estimation of total and peak rainfall runoff and sediment load are discussed in chapter four while fifth chapter deals with water storage reservoirs and spillways. In sixth chapter land levelling and earthwork estimation are discussed. The key sector of watershed that is soil conservation against water erosion, is discussed in the seventh chapter. The dryfarming technology, relevant for different agroclimatic conditions of India is thoroughly discussed in chapter eight. An exclusive chapter nine has been devoted to efficient use of available rain and ground water resources of the watershed area. The most important aspect of watershed-monitoring and evaluation is described in chapter tenth. As an illustration of watershed management, a case study of Bunga Watershed has been presented in chapter eleven. The latest technology of remote sensing and GIS applicable to watershed has been discussed in chapter twelve. The entire text is presented in a simple and lucid language with several numerical problems and their solutions besides illustrations. It is believed that this book will serve as an important study tool on watershed for undergraduate and postgraduate students, teachers, agricultural development staff and others involved in watershed management programme.

I express my sincere thanks to all those who helped me in any way in preparing the manuscript of this book and to M/s Associated Publishing Company, New Delhi, for critical review of this manuscript and publishing it in a record time of three months.

M. C. Oswal

Contents

Chapter — 1

Introduction

Dryland Agriculture

Dryland agriculture is the mainstay of over 76 percent inhabitants in the developing countries of the world most of which are located in the tropics (region of the earth situated between 23.5^{0} north and south of equator). A majority of those inhabitants rely on traditional agricultural systems which use fragile soils and depend on undependable rainfall as the only source of water available for crop production. Of the total potential arable land of 1843 Mha in 90 developing countries, 662 Mha (36%) were rainfed compared with 132 Mha (7.5%) that were irrigated. Till eighties there was substantial increase in irrigated land but the rate of increase in irrigated area declined afterwards due to financial constraints and limited availability of water. Therefore, it is becoming increasingly important that greater proportion of food, fibre and fuel needs of inhabitants of the developing countries got to be provided from dryland areas.

India has a geographical area of 329 million hectares and drylands which are either rainfed or under limited irrigation, constitute about two-third of the net sown area of 143 Mha. These areas account for about 40 per cent of cereals, 70 per cent of cotton and oilseeds and 90 per cent of pulses. The scope to expand irrigation being limited due to physical, climatological and economic factors, it is obvious that further increase in the annual food grain production, to about 240 million tonnes, needed by the country by the turn of the century, will have to come from the drylands.

It is well recognized that drylands are not only thirsty but hungry as well and are cultivated by the poor sections of population. These areas produce much below their capacity. Factors such as variable climate and soil quality, high population pressure, traditional agricultural technologies and limited access to capital together have resulted in low production of dryland crops. The degradation of environment in the dry land areas is basically attributed to the increasing biotic pressure on the fragile ecosystems in the absence of adequate investments and appropriate management practices to augment and conserve the land and water resources. Population growth and poverty on the one hand and the pressure of rising demand from affluent on the other, have been exerting powerful pressure on the ecosystems. The macro-economic policies which provide inducement to the over exploitation of natural resources are also responsible for denudation of environment. For example in the dryland areas, the pumping of water has been proceeding at a faster rate than the rate at which ground water is being recharged. This is on account of the availability of electricity at a flat rate regardless of the amount of electricity used for pumping water. The break down of traditional institutions to individualized interest for managing common property resources and the failure of new institutions to fill the vacuum have also been responsible for the denudation of natural resources.

Development of dryland agriculture is essential for national food security, narrowing down regional imbalance, creating rural employment and maintaining favourable ecological balance in tropical countries. The productivity of the dryland is not only low (< 0.5 t ha^{-1}) but is also highly unstable and has remained stagnant despite several research efforts. This is primarily attributed to frequent failure of crops due to severe moisture stress at one or more of their critical stages of growth. Availability of small quantity of water at these critical stages can dramatically increase crop production, besides building confidence of the farmers in dry farming.

Soil Conservation Programmes

Water is the most limiting factor for crop production and water management is the key to development of sustainable agriculture for rainfed areas. Land degradation due to soil erosion by water resulting in loss of top fertile soil is another most limiting factor responsible for low yields of dryland crops. The policies for water management in dryland areas should therefore be so oriented that it also helps dryland farmers to take up soil conservation measures.

In India efforts to adopt soil water conservation measures have been promoted largely through central government sponsored programmes implemented by the relevent state departments and field agencies. Some of the important programmes taken up in the past have included:

Soil Conservation in the Catchment of River Valley Projects

This programme was initiated in the Third Five Year Plan period (1961-66) following the concern brought out by several surveys revealing that the siltation rate of most reservoirs was much higher as compared to the rates assumed at the time of planning, which could adversely affect the planned life of these reservoirs. Initially the programme was taken up in 13 catchments and later was extended to include 27 river valley project catchments covering a total area of about 69 Mha. For implementation of these programmes, the central government provided 100 per cent assistance comprising 50 per cent grant and 50 per cent as loan to the concerned state government. The actual field execution is carried out by the state department such as Agriculture, Soil Conservation and Forests. Till the end of 1988-89, nearly 2.27 Mha were treated at an expenditure of 2729 million rupees. Although the area treated annually has remained almost constant over the past one decade, the unit costs have been increasing sharply (Subramaniyam and Samuel, 1990).

In addition to the specific programme for soil and water conservation, the Government had sponsored several programmes for rural poor, particularly to benefit areas with low and uncertain rainfall. Most of these programmes included a component relating to conservation and upgradation of the resource base. For example, in 1970 a Rural Works Programme was initiated in 54 districts of 13 states concentrated in the western and central zones of the country. Although the intent of the programme was to create permanent productive assets, almost from the inception of the scheme, a very substantial share of the total funds available, was used to provide employment in the creation of large number of works of low priority with little or no productive or development impact.

In order to correct the situation, this programme was replaced in 1973 by the Drought Prone Area Programme (DPAP), the intent being to change from a relief and employment oriented programme to one aimed at drought proofing. Major attention was to be placed on restoration of ecological balance, soil and water conservation, afforestation, restricting cropping patterns and improvement of agronomic practices. In 1980-85 various soil and water conservation treatments were provided to an aggregate area of 0.42 Mha. However, doubts were frequently expressed on the validity of achievements in terms of reported area coverage and the impact of soil conservation engineering works. This is also true for other rural development programmes e.g. Desert Development Programme started in the Seventh Plan period (1985-86) covering 21 districts in 5 states over an area of 36.2 Mha; the National Rural Employment Programme, etc. According to available information (Abrol and Sehgal, 1994) till the year 1987-88, in about 33 Mha area, soil conservation measures had been taken but little information is available about the impact that these measures helped in stabilizing the resource base.

Lessons from Past Efforts

In India, soil and water conservation measures, recommended and carried out at the field level, have been largely at the instance of Central and State Governments. Employment generation and mitigation of drought conditions were the key elements intended to be achieved. At the field level, the most common soil conservation technique used was the constructions of contour or graded bunds at a vertical interval of 1 to 2 m on slopes of 6–8 per cent or less. Experience has shown that small, resource-poor farmers are often reluctant to realign their plot boundaries to confirm to the acceptable contour grades. Farmers are also reluctant to have these bunds within their field since these take a considerable chunk of their land out of cultivation. For these and other reasons, the technical considerations are often compromised. As a result in the event of high rainfall, the bunds frequently breached and the purpose for which these were constructed was rarely achieved. Often erosion on these fields accelerated due to soil disturbance. Recent studies have shown that on black soils contour bunding was not beneficial irrespective of rainfall and in areas with more than 600 mm rainfall irrespective of the soil type (Bhumbla, 1994).

Many of the past conservation efforts were carried out without involving the land owners in an effective manner. Infact, there were complete lack of mechanism to interact with the farmers leaving them only as silent spectators. Yet another constrain in some of the past soil and water conservation programmes was their complete isolation from the crop production programmes. Thus while seeds and fertilizers were considered an input for improved productivity, soil and water conservation were not. For this reason, the farmers were not enthused from these programmes. A small and resource-poor farmer is more interested in improving production of his food grain and income and not in the long term sustainability. Unless he is convinced that soil and water conservation measures lead to his production, he is not going to accept them. Further, such programmes were taken up only in public arable lands. Therefore, they could not attract much attention of farmers.

It is apparent that many of the government sponsored activities ultimately became instrumental to provide temporary employment to the rural unemployed rather than being a meaningful programme to develop crop land and raise its productivity.

Watershed Approach

Notwithstanding past setbacks, there has been welcome qualitative changes in our approaches to the management of our resources in the past decade or so. It has been increasingly realized that, in future, there is a need to adopt an integrated approach towards the management and development of our resources including soil, water, crop, range, forest and livestock on the basis of natural ecological unit after involving the people of the area.

Since 1983-84, the Indian Council of Agricultural Research working through the Central Institute for Dryland Agriculture, Hyderabad and the Central Soil and Water Conservation Research and Training Institute, Dehradun and collaborating closely with the State Agricultural Universities and line departments launched efforts towards improved management of 47 experimental model watersheds in different agro-climatic zones for intensive development and for in depth monitoring and evaluation. In 1986-87, a National Watershed Development Programme for Rainfed Agriculture was initiated with a primary objective of stabilizing agricultural production in rainfed areas by significantly increasing investment in these areas. The programmes addressed to rainfed areas falling in the zones 500-1125 mm and above 1125 mm per annum. The main objectives of the programmes were

(i) Taking watershed as a basis, conserve and up grade crop lands and degraded lands.
(ii) Develop and demonstrate location specific technologies for soil and water conservation and for crop production stabilization.
(iii) Augment the fodder, fruit and fuel resources of village communities through appropriate alternate land use systems.

In the last decade, there have been outstanding examples of success which show that drought can be beaten, land can be protected, water can be conserved and productivity and income of the people in these areas can be improved provided concerted efforts on watershed basis are made, with motivated and dedicated leadership from the Administrators and involvement of voluntary organizations together with participation of local people. There are outstanding examples of success at Ralegeon Sidhi and Adgaon in Maharashtra, Kabbalnala and Mittemari in Karnataka, Jhabua in Madhya Pradesh and Bunga and Shukhomajri in Haryana, all of which show that drought can be beaten.

Therefore in recent years, there has been a marked increase of both the level of interest in this approach and also in the willingness to commit resources for implementation. Consequently, a number of watershed Development Projects have been taken up under different programmes viz (i) Drought Prone Area Programme (ii) Desert Development Programme (DDP) and (iii) Integrated Watershed Development Projects Scheme (IWDP) of the National Watershed Development Board. So far these programmes have laid down their own separate guide lines, norms, funding patterns and technical components based on their respective and specific aims. While the DDP focussed on reforestation to arrest the growth of hot and cold deserts, the DPAP concentrated on non-arable lands and drainage lines for *in-situ* soil and water conservation, agro-forestry, pasture development, horticulture and alternate land uses. The IWDP, on the other hand made silvipasture, soil and moisture conservation on wastelands under government/community/private control as their predominant activity. The National Watershed Development Programme for Rural Areas (NWDPRA) combines the features of all these three programmes with the additional dimension on improving arable lands through better crop management technologies. While the focus of these programmes may have differed, the common theme amongst these programmes has been their basic objective of land and water resources management for sustainable production.

On the recommendation of the Technical Committee on DPAP and DPP (1994) headed by Dr. C. H. Hanumantha Rao, these programmes have been integrated under the Ministry of Rural Development with a common set of guidelines with regards to objectives, strategies for implementation and expenditure norms.

Definition, Scope and Objectives of Watershed

A watershed strictly refers to divide/separating one drainage basin from another. Hydrologically a watershed is an area of land, the runoff from which drains to a particular point in a drainage system. In other words, the land area from which water drains to a common point is a watershed. It is thus a self defined area which does not allow any water from outside the catchment to enter it and allows its own water to discharge to a common point in a stream. Accordingly, a watershed (catchment) must have a ridge line at elevation higher than its surrounding, inner as well as outer areas. A watershed therefore can be defined as an area of land that contributes runoff to a commom point and is separated from adjoining areas by a natural elevation ridge-line.

Dryland Agriculture on watershed basis ensures efficient use of available water (both rainfall and surface and underground waters) and other natural resources like fauna, flora, cattle, soil, climate and human beings for the ultimate benefit of the mankind without disturbing the natural ecosystem. Therefore, the programme requires an intensive and integrated approach including crop production, horticulture, forestry, pasture development, water harvesting, moisture conservation, use of improved seeds, implements, fertilizers, pesticides etc. besides animal husbandry, poultry, piggery and rural development by a multi-disciplinary team comprising of agronomist, forester, soil scientist, agricultural engineer, horticulturist, sociologist, economist, veterinarian, home scientist, planner, executor, farmers etc. Watershed management involves a combination of practices for the overall development of the watershed area. This development recognizes watershed as a boundary irrespective of its administrative and other boundaries.

Among tropical countries, India could be counted among a few countries which have undertaken dryland watershed development programmes in a big way. The basic principle in watershed development has been to retain the precipitation within the watershed, reduce top soil loss and minimise runoff. This has been achieved through gully plugging, check dams, contour ditches, percolation tanks, water reservoirs/ponds etc; depending upon the terrain. These measures, in addition to making surface water available for some period for crop utilization, also help in recharging the aquifers for using ground water, later on, for irrigation. A further benefit is soil conservation water helps in increasing land productivity and in reducing silt flow from such lands into streams and rivers and finally into reservoirs or seas. Though an integrated planning approach and adoption of modern technologies for dryfarming, biomass (grain, fodder, fuel, etc.) production has increased, but the development has not been sustainable in terms of operation and maintenance of assets, created and common property resources because of inadequate participation of the village communities and user groups. In practice the real integration of various activities at the field level could not be made. According to a survey, carried out by the Agricultural Finance Corporation (in 1988), in one of the schemes "the line departments separately prepared their part of the plan and the two are superficially combined and called watershed management plan. Also, relative to the size of problem and the amount of resources committed to these activities, integrated watershed development programmes so far, remained in their infancy. To overcome these shortcomings, National Watershed Development Programme for Rural Areas (NWDPRA) has been launched since April, 1995 with the following objectives:

Objectives of Watershed

1. To promote the economic development of the village community which is directly or indirectly dependent on the watershed through:
   (a) optimum utilization of the watershed's natural resources like land, water, vegetation etc. that will mitigate the adverse effects of drought and prevent further ecological degradation.
   (b) employment generation and develoment of the human and other economic resources of the village in order to promote savings and other income-generating activities.
2. To encourage restoration of ecological balance in the village through:
   (a) sustained community action for the operation and maintenance of assets created and further development of the potential of the natural resources in the watershed.
   (b) simple, easy and affordable technological solutions and institutional arrangements that made use of and build upon local technical knowledge and available materials.
3. Special emphasis to improve the economic and social condition of the resource-poor and the disadvantaged sections of the watershed community such as assetless and women through:
   (a) more equitable distribution of the benefits and land and water resources development and the consequent biomass production.
   (b) greater access to income generating opportunities and focus on their human resource development.

The guidelines of the National Watershed Development Project for Rainfed Areas (NWDPRA) in India emphasise the Government Policy resolution to stimulate and promote people participation in project planning, project preparation and implementation and post project management of project assets as an integral part of the approach and strategy for watershed development. The NWDPRA has been designed to operate either through the established democratic decentralised institutions like the village Panchayat or promote development of specific organisations to implement the programme. The basic objective behind people's participation is the ultimate goal to convert the watershed development project from a government scheme to a people's movement and where the government participation is only to necessary support. Institutionalising people's movement is feasible only through organisation of self regulating beneficiary groups.

The framework of NWDPRA reads as follows. All works in NWDPRA would be implemented through the beneficiaries themselves on common lands by the village panchayat or a group or association of beneficiaries residing in the micro-watershed. No contractors are to be engaged. For initiating greening of common lands they should not be closed in their entirety in the initial year. Such a step is equivalent to initiating conflict between the project and the people. Instead 1/10th of the common area should be enclosed in the first year and every year thereafter. Within a few years the benefits of cut and carry are established in the minds of the villagers and the process of stall feeding will be initiated. With this, the natural tendency to have fewer cattle of improved breed would set in. The guidelines also provide for selection of a landless, agricultural labour in order to signify his or her concern for tending cattle as well as to look after the common grazing lands. The labour would focus attention on the natural resource management on behalf of the village community. The approach of NWDPRA for implementation of the programme was adopted in the staff appraisal report of the Integrated Watershed Development Programme also.

The programme of NWDPRA are designed to be executed by duly elected people's organisations called User Committees. Each User Committee comprises of 7 to 11 members depending on the population of the village with representation for women, weaker sections, the scheduled castes and scheduled tribes. Each User Committee is headed by a Chairman indirectly elected by the members of the User Committee itself. The function of the User Committee includes formulation of the watershed development plan, design and implementation of the plan by the beneficiaries themselves or by employing landless agricultural labour, responsibility for generated usufructs and maintenance of generated assets. The User Committee is assisted by a member secretary who is a Junior Engineer of the Watershed. The User Committee shall also maintain all the records prescribed by the State Government and place all the accounts before the Gram Panchayat. Cultivators' contribution in the cost of various development works is introduced to promote participatory development. This contribution varies from 15 to 40 per cent depending on the nature of work and interest of the farmers.

In the NWDPRA attempts have been made to integrate five programmes viz; DPAP, DDP, IWDP, IJRY (Indira-Jawahar Rojgar Yojna) and EAS (Employment Assurance Scheme) to overcome several administrative problems and pulling 50 per cent financial resources of these programmes exclusively for watershed development on village basis. Thus, a village watershed would be covered either under DPAP/DDP/IJRY/IWDP/ EAS for its source of funding (50%) for taking up all the activities envisaged in the watershed development projects.

Chapter — 2

Climate and Soils

Soil-water conservation measures and practices largely centre around characteristics of climate and the nature and properties of soils of a region. Basic information on these parameters is vital for planning various measures aimed at efficient conservation of soil and water resources.

Arid and semi-arid areas, often called drylands, are predominant in tropics, where crops are either rainfed or under limited irrigation. The climatic factors which exercise major control in soil management and agricultural production are rainfall and potential evaporation. In dryland agro-ecosystem, climatic variability is the major cause of inter-annual variability of crop yields. These water deficit environments are unfavourable for the growth and development of rainfed crops and often produce low and unstable crop yields (Singh and Subha Reddy 1988). The effective cropping season is restricted both by quantum and distribution of rainfall, thereby setting limits on the choice of crops, cultivars and cropping system. For post-rainy season crops, usually grown on conserved soil moisture, it is the moisture storage at sowing time that determines the choice of crops and cultivars. Crops, cultivars, and cropping system should be selected as such that their growth characteristics fit well into the period of moisture availability. This shows that land use planning, crop zonation and other agricultural activities require specific agro-climatic information on factors such as the beginning, ending and duration of the wet or growing period and among other things the probability of dry spells during the growing season.

Rainfall Regions of India

Rainfall is the main source of both surface and underground water. All assessments of availability of stream river or ground water are based on rainfall with some assumptions about run-off and recharge of ground water. Rainfall in India (Fig. 2.1) varies from about 200 mm in Jaisalmer district of Rajasthan to more than 3000 mm in Western Ghats and parts of North-Eastern states. The crop water demand is governed by potential evpotranspiration of the area.

It is defined as evapotranspiration of a thick young grass under sufficient water supply condition. The potential evapotranspiration (modified Penman), as determined by India Meteorological Department, for different regions varies from less than 750 to more than 2000 mm. India can be divided into following rainfall regions:

(i) very low rainfall region (average annual rainfall less than 500 mm)
(ii) low rainfall region (average annual rainfall 500-750 mm)
(iii) medium rainfall region (average annual rainfall 750-1150 mm), and
(iv) high rainfall region (average annual rainfall more than 1150 mm).

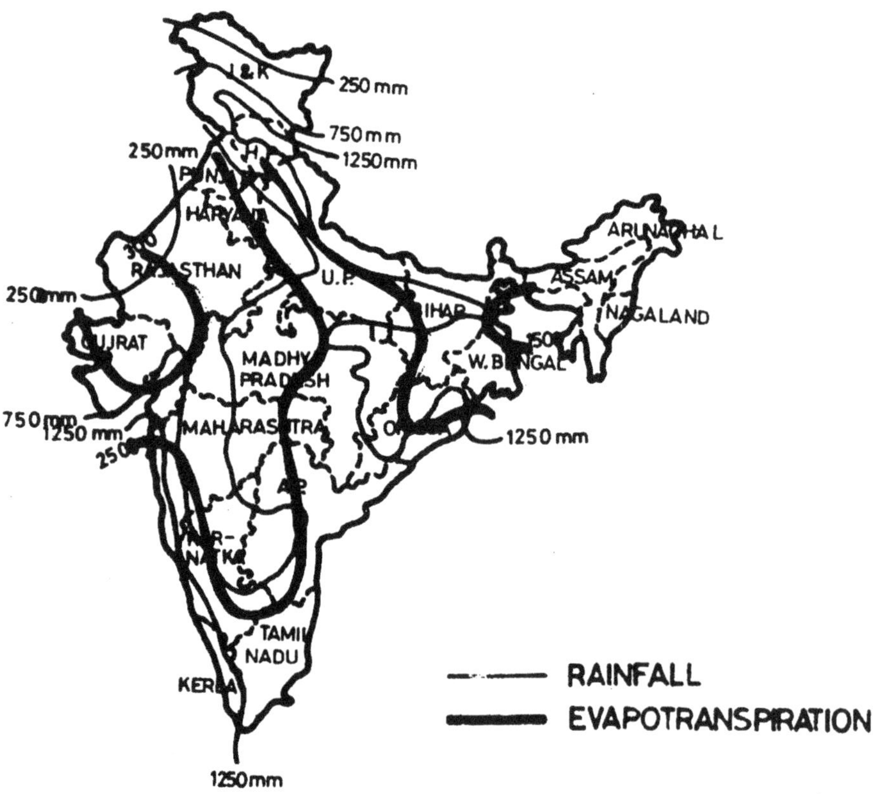

Fig. 2.1: Distribution of Rainfall and Potential Evaporation in India

Based on the normal rainfall (1901-1950) as published by Department of Meteorology and taking district as a unit the total reporting area of very low, low, medium and high rainfall regions has been estimated to be 30.0, 55.8, 90.4 and 100.2 Mha, respectively. The Himalyan states along with U.P. Hills with total reporting area of 27.9 Mha has been accounted so far because of wide variations in topography, rainfall, soils and land use. The very low rainfall districts are confined to Western Rajasthan (2/3 of the total very low rainfall area), Kutch and Jamnagar districts of Gujarat, West Haryana and West Punjab, the rainfall is not only low but highly variable from year to year. Crops can be raised only in the monsoon season and that too of short durations (pearl millet, mung bean, kidney bean, clusterbean). About 50 per cent of total pearl millet area and 90 per cent of mothbean and clusterbean together occupy about 10 Mha. Crops failure is common and yields range from a few kilogram to about 500 kg/ha. The area may face floods once in 10 years. In Rajasthan cultivation of crops has spread to marginal lands. The ground water is the main source of irrigation in Rajasthan except Ganganagar district but it is inadequate and getting depleted at a faster rate. The priority in rain water management in this area should be to develop grassland and bush/shrub land.

The low rainfall areas comprises of Punjab, Haryana, Western Uttar Pradesh, Eastern Rajasthan and North Gujarat, West Maharashtra, North Karnataka and Royalseema area of Andhra Pradesh. In the alluvial region of Indo-gengetic plains ground water is the main source of irrigation and cropping is dependent on tube wells. Canal irrigation is confined to West Uttar Pradesh, where sugarcane is the main crop. In the shallow and medium black soil area, sorghum, pearlmillet, groundnut and pulses are the main rainfed crops and the yields are low. Most of the groundnut in the country is grown in this region but in years when monsoon withdraws early the yields are reduced drastically. Even in this area, ground water is the main source of irrigation which is inadequate in most of the years. Recharge of ground water should receive highest priority. Soil conservation and dryland practices are important. The area under cultivation should be reduced and diverted for low water requiring crops. Collection and storage of excess rain water in farm pond for providing life saving irrigation and to recharge ground water are essential.

The medium rainfall area comprises of three submountainous districts of Punjab and Haryana, most of Uttar Pradesh, part of East Rajasthan, Vidharbha region of Maharashtra, Central Gujarat, South Karnataka, most of Andhra Pradesh and Tamilnadu. The crops of the region are maize, sorghum, finger millet and cotton during *kharif* (summer) and wheat, gram mustard, pulses during *rabi* (winter) on conserved moisture. The region faces the problem of both excess and lack of water. This is the area where rain water management to recharge the ground water should have highest priority. Soil conservation to retain as much rainfall as possible in the field without harming the crop and drainage of excess rain-water through appropriate grassed water ways and its storage in depressions should be the main programmes. In the hard rock red soil area of Jhansi district sinking of dug wells helped in increasing the area of ground-nut which has entirely replaced sorghum because the farmers have access to water in the event of failure of rain to provide irrigation. Area and yield of wheat could be increased if wells had enough water up to end of

March. The area has the maximum scope of both surface reservoirs and recharge ground water.

The high rainfall areas cover about one-third of the country in parts of East Uttar Pradesh, Bihar, West Bengal, Orissa, Assam, East Madhya Pradesh, Kerala, Coastal Karnataka, Konkan region of Maharashtra and South Gujarat. Rice occupies more than 90 per cent of the cropped area in this region during rainy (*kharif*) season which is grown under upland, lowland and deep water situations. Floods are common. but lack of rains at the time of flowering results in a complete failure of the crop. The total rain even in the worst year is sufficient if it is uniformly distributed. Surface water storage in natural depressions/brooks should get priorities in the development works to contain floods and provide drainage and irrigation. In the winter season cropping in absence of irrigation on residual moisture is confined to pulses and oilseeds. Development of ground water should receive highest priority to harness good yields of winter crops. In West Bengal, winter rice produced about 3 t ha^{-1} of rice where ground water is used for irrigation.

Soil Resources of India

Soils of India broadly (Fig. 2.2) fall into five major soil groups viz., alluvium (derived by water), red, black, desertic and sub-mountainous. The estimated area of these broad groups is given bellow:—

(i) Alluvium (Combusols) soils
  Inland - 63.1 Mha
  Deltaic - 6.8 "
  Coastal - 4.4 "

(ii) Black soils (Vertisols) - 73.3 Mha

(iii) Red, yellow and laterite (Lixasols and Nitisols) - 87.6 Mha

(iv) Desertic (Aridosols) - 30.3 Mha

(v) Submountanous (Leposols) - 28.7 Mha

The important characteristics of these soils are as follows:

Alluvial Soils: The texture of these soils may vary from loamy sand to silt loam. Usually they have low clay content, high sand content and possess low (10-12%) water and nutrient holding capacity and high water intake rate (> 25 mm hr^{-1}) but limited soil depth. They develop surface crusts with a beating rain. When slopes occur, in the lower parts of the slope, the soils may be saline and heavier in texture. They are usually neutral to alkaline in soil reaction, low in nitrogen, low to medium in phosphorus and medium to high in potassium. These soils are usually good for cultivation.

Geographically, the alluvium is divided in to *khadar* (the newer alluvium with less *kankar*) and *bhangar* (the older alluvium with full of *kankar*). These soils may have hard pan at shallow depths due to continuous ploughing with a local plough for centuries. Deep ploughing, inter-culturing to break soil crusts and application of balanced dose of fertilizer are required for maintaining their productivity.

Black Soils: These soils are usually heavier in texture with clay contents varying from 30-70 per cent and possess high water holding capacity, low infiltration rates and poor aeration, high water run off potential (30-40%) and soil erosion problem and narrow tillage range moisture and develop deep cracks on drying. Their depths may vary from low to medium to very deep. They are rich in bases, poor in nitrogen, phosphorus and organic matter and have high reserve of K, Ca, Mg, S (secondary nutrient). Elements like boron may reach toxic levels in the salt affected soils. Salinity in sub-soils and presence of

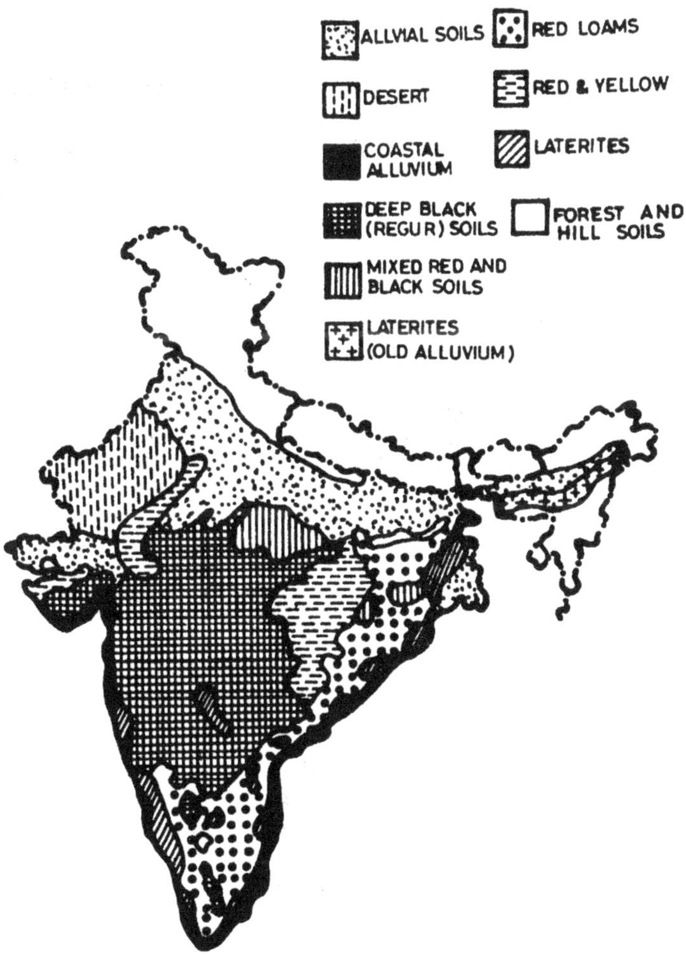

Fig. 2.2: Soils of India

calcium carbonate concretion due to poor drainage are common. Timely deep tillage and application of vegetative mulch besides balanced dose of fertilizer and proper bunding and levelling of fields are the measures required to keep them fit for cultivation.

Red Soils: These soils are light to medium in texture and are usually low in clay contents (10-20%). They are shallow to medium in depths and develop surface crusting and compaction of the top soil after rains. They possess low nutrient holding capacity (CEC, 10 me/100 g) and are universally poor in nitrogen, usually in phosphorus and occasionally in potassium. On continuous cropping calcium becomes deficient in soils. Similarly in high rainfall areas sulphur is found deficient. Deep ploughing, inter-culturing and application of fertilizer are required to maintain their good health.

Desertic Soils: These soils are very light in texture and possess very low water rentention capacity, high infiltration rate (> 25 mmh^{-1}) and prone to wind/water erosion due to low clay and organic matter contents. The presence of calcium carbonate concretion at varying depths and sub-soil salinity is also common in these soils. Medium textured desertic soils develop surface crusts after beating rains. They are low in nitrogen, medium in phosphorus and high in potassium.

Sub-mountainous Soils: Due to slopy lands, the soil depth is a limiting factor in these soils. Land slides and soil erosion by water is a common problem interfering with the normal cultivation of these soils. They usually possess high water holding capacity and are usually low to medium in nitrogen and phosphorus and medium to high in potassium. Proper bunding and other soil conservation measures, suitable crop planning and fertilizers application are the measures required for their agricultural use.

Geographical Distribution and Bio-climatic Characterization of Drylands

According to the degree of aridity index (defined as the rainfall deficit expressed as fraction of potential evapotranspiration on annual basis), the Indian territory may be divided into four categories: (i) arid lands covering the north-western portions of the Thar desert bordering Pakistan, 5% of the overall dry areas of the country (ii) semi-arid lands adjoining these arids (25%), sub-humid (dry) lands (30%) and transitional lands with adequate moisture supply (40%). An area is termed arid, semi-arid, dry-subhumid and humid, respectively, if the aridity index is greater than 0.66, between 0.66 to 0.33 and 0.33 to 0.00 and less than 0.00. The most arid region of the Thar Desert in the north-western part, covers a greater part of Rajasthan and smaller parts of the adjoining states of Haryana and Punjab. Desertic and semi-desertic conditions also prevail in Rann of Kutch and Kathiawar peninsula and in most parts of Gujarat state. Aridity tends to decrease in north-eastern direction. However, the zone of unstable moisture stretches across a rather wide belt in peninsular part of India covering states of Maharashtra, Madhya Pradesh, Andhra Pradesh, Karnataka and Tamilnadu.

The arid and semi-arid territories correspond with the zones of subtropical and tropical climate that extend, with a few exceptions, along the tropics in the northern and southern hemisphere. According to Koppen (1931), who used vegetation concept in which the aridity based on precipitation and temperature, desert vegetation is the expression of the arid climate while the primary thorn forest is indicative of semi-arid climate. Climatic analysis based on long-term averages of temperature and precipitation are found to be of limited use, since

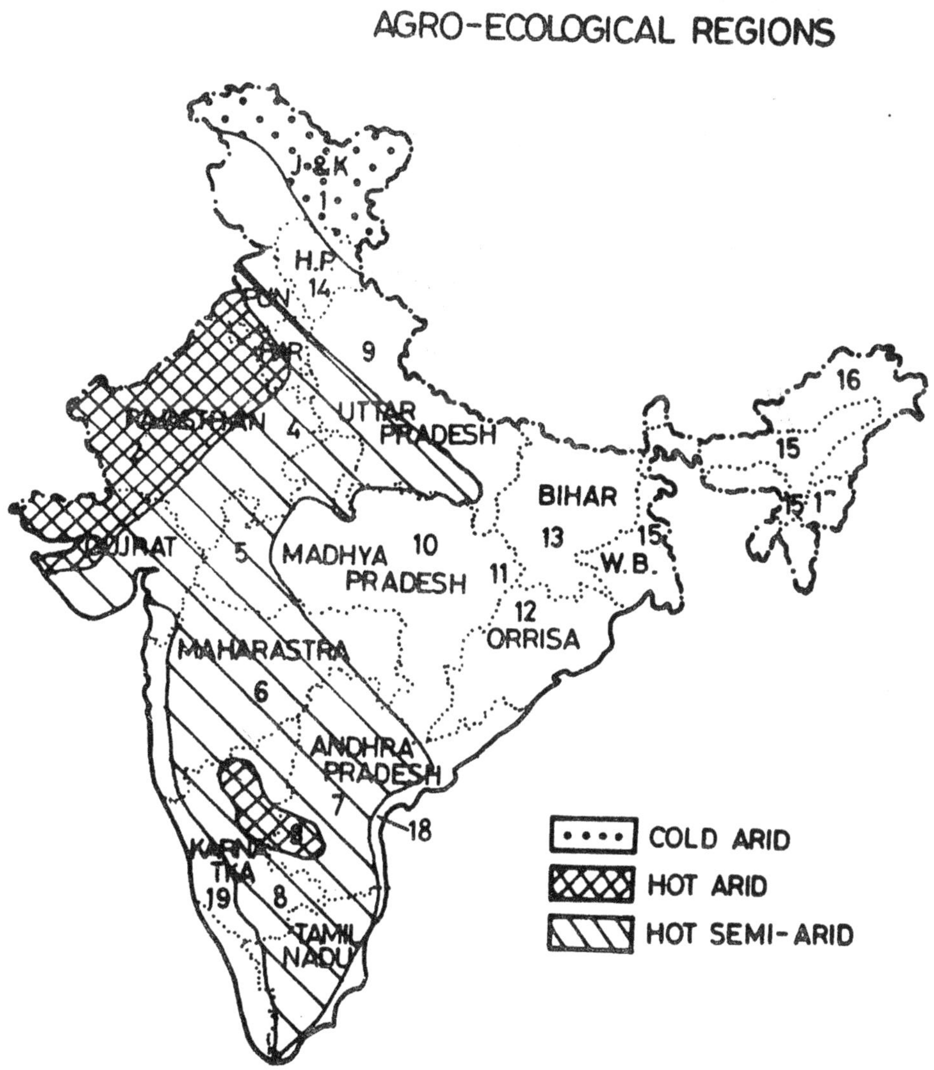

Fig. 2.3: Bioclimatic Zones of India

they mask the effect of weather variability and often do not relate directly to agricultural problems. Meher-Homji (1975) suggested that species area statistics used in combination with climatic characteristics provide a better indication of the bioclimatic degree of aridity of a region than the climatic data alone. Bhattacharjee *et al.* (1982), based on aridity, humidity, and moisture indices of Thonthwaite (1948) and the thermal efficiency, established bio-climatic types of India to facilitate delineation of the arid and semi-arid areas in India. Correlating the bioclimates and growing periods, it has been observed (Sehgal *et al.*, 1992) that the arid region corresponds with the growing period of less than 90 days while the semi-arid region corresponds to that of 90-150 days. Taking into account the important climatic parameters like precipitation (P), temperature (T), potential evapotranspiration (PET) and using various indices, the salient characteristics of arid and semi-arid bioclimates of India are given below.

Arid bioclimate (Fig. 2.3) encompasses extensive parts of India and comprises arid cold and arid hot climate based on thermal regime. The arid cold with no water surplus occupies Ladakh region of Jammu and Kashmir and is characterized by mesothermal regime, mild summers and severe winters with mean annual temperature, less than 8°C and mean annual rainfall of less than 150 mm. The precipitation covers less than 15 per cent of annual PET. The area qualifies for aridic soils moisture regime and cryic soils temperature region with a short annual growing period (GP) of less than 90 days. The arid hot area with little or no water surplus occupies western and north-western areas of Rajasthan, Gujarat and Punjab bordering Pakistan and western part of Haryana state. It is characterized by hot summer and cold winter; rainfall ranging between 300-400 mm; potential evapotranspiration exceeding 1900 mm in most of the years. Except for 2-3 months of uncertain showers of high intensity, rest of the time in most of the years is dry with severe drought leading to high water deficiency ranging between 1200-1400 mm annually having no surplus. The area represents aridic soil moisture regime and hyperthermic soil temperature regime with an annual growing period of less than 90 days.

Semi-arid bioclimate covers major part of the central and south-central peninsula extending to western and north-western parts of the country. The climate of the region is characterized by hot and dry summers and cool winters. The annual precipitation ranges from 400-750 mm covering 40 per cent of the mean annual PET which ranges between 1400-1900 mm. The area shows an annual water deficit of 700-1000 mm. The moisture regime is ustic, while the soil temperature regime is hyperthermic.

Arid and semi-arid agro-ecoregions in India and their major constraints are reported in table 1.

Table 2.1: Agro-ecoregions in India and their microconstraints

Ecoregion	*Coverage*	*Area Mha*	*Constraints*
Arid Ecosystem	i) Western Himalayas, cold arid region with shallow soils	15.6	Shallow, light textured, and highly calcareous gravelly soils, low temperatures and scanty rains.
	ii) Western plain, Kutch and part of Kathiawar peninsula, hot arid region with desert and saline soils.	29.6	Erratic and scanty rainfall, high PET, light textured, saline/sodic soils deficient in N, Zn.
	iii) Deccan plateau, hot arid region with red and black soils.	4.7	Poor drainage, high run-off and erosion, long dry spells.
Semi-arid Ecosystem			
	i) Northern plain and central highlands and Aravallis hot region with alluvium (derived) soils	32.9	Light soils with low water hodling capacity and over exploitation of ground water, low in N and P.
	ii) Central (Malva) highlands, Gujarat plains, Kathiawar peninsula hot region with black soils	18.4	Poor drainage, high runoff and erosion.
	iii) Deccan-plateau hot region with shallow to medium black soils	33.0	Salinity/alkalinity problem in irrigated areas, low soils depths, high runoff, poor drainage, erosion.
	iv) Deccan (Telangana) pleateau and Eastern ghats, hot region with red and black soils.	20.8	High runoff and soil erosion, salinity, low in N, P, Zn.
	v) Eastern Ghats, Tamilnadu uplands, Deccan (Karnataka) Pleateau hot region with red soils.	22.7	High runoff and soil erosion, salinity, low in N, P, Zn.

Growing Season

The length of the growing season can be expressed as the sum of the length of rainy season and the period for which the water stored in the soil at the end of rainy season and the post rainy season rainfall if any can meet the water requirement of arable crops. The length of growing season can be estimated by using water balance model by computing (PET) and the actual evapotranspiration (AET) for a given water holding capacity of the soil (Raman and Srinivashan Murthy, 1971). The ratio of AET/PET is referred to as the index of moisture adequacy (IMA). The type of water availability is classified as the basis of IMA value as given in Table 2.2.

Table 2.2: Types of water availability on the basis of IMA value

Value of IMA	*Type of Water availability*
0.00 to 0.25	Dry
0.25 to 0.50	Semi-dry
0.50 to 0.75	Sub-moist
0.75 to 1.00	Moist
1.00	Humid

The length of growing season is assumed to commence from the period when IMA is 0.50 or more and extends up to the period during which IMA value does not fall below 0.25. In this method it is stressed that (i) the water availability for AET is equal to PET during the period when the rainfall is equal to or greater than the PET (ii) excess rainfall over PET contributes to the soil moisture storage and (iii) runoff and deep drainage will occur after the soil reaches the field capacity moisture level. The length of growing season computed for some of the locations in India in the dry farming belts assuring that the AET/PET equal to 0.25 and continued to be more than 0.25 thereafter, is given in Table 2.1.

Table 2.3: Growing season at different locations

Location	*Mean Annual Rainfall, mm*	*Length of Growing season weeks*	*Duration*
Agra	589	24	2 July - 16 Dec.
Akola	876	27	11 June - 16 Dec.
Anand	626	25	18 June - 9 Dec.
Anantapur	543	13	13 Aug. - 11 Nov.
Bellary	518	8	16 Sept. - 8 Nov.
Bhubneswar	1525	31	3 June - 31 Dec.
Bijapur	658	17	16 Sept. - 8 Nov.
Dehradun	2324	51	4 June - 27 May
Hebbal	819	32	14 May - 13 Dec.
Hisar	420	13	2 July - 30 Sept.
Hyderabad	764	33	18 June - 18 Nov.
Indore	963	36	18 June - 18 Feb.
Jhansi	885	21	25 June - 18 Nov.
Jodhpur	380	11	2 July - 16 Sept.
Kovilpatti	792	21	3 Sept. - 11 Feb.
Ludhiana	531	25	25 June - 25 Feb.
Rajkot	626	17	25 June - 21 Oct.
Ranchi	1463	45	2 June - 8 April
Rewa	1109	36	18 June - 25 Feb.
Samba	1081	44	25 June - 29 Nov.
Sholapur	742	23	4 June - 11 Nov.
Udaipur	709	22	25 June - 25 Nov.
Varanasi	1113	32	25 June - 4 Feb.

The growing season concept has following limitations:

(i) Large variability in the commencement and length of growing season from year to year, and

(ii) It depends upon the available water holding capacity of the soil and increases with increase in water holding capacity.

Possible Cropping Intensity

The cropping intensity in the dryland areas is usually 100 per cent, implying that only a single crop (rainy season crops/post rainy season crops on

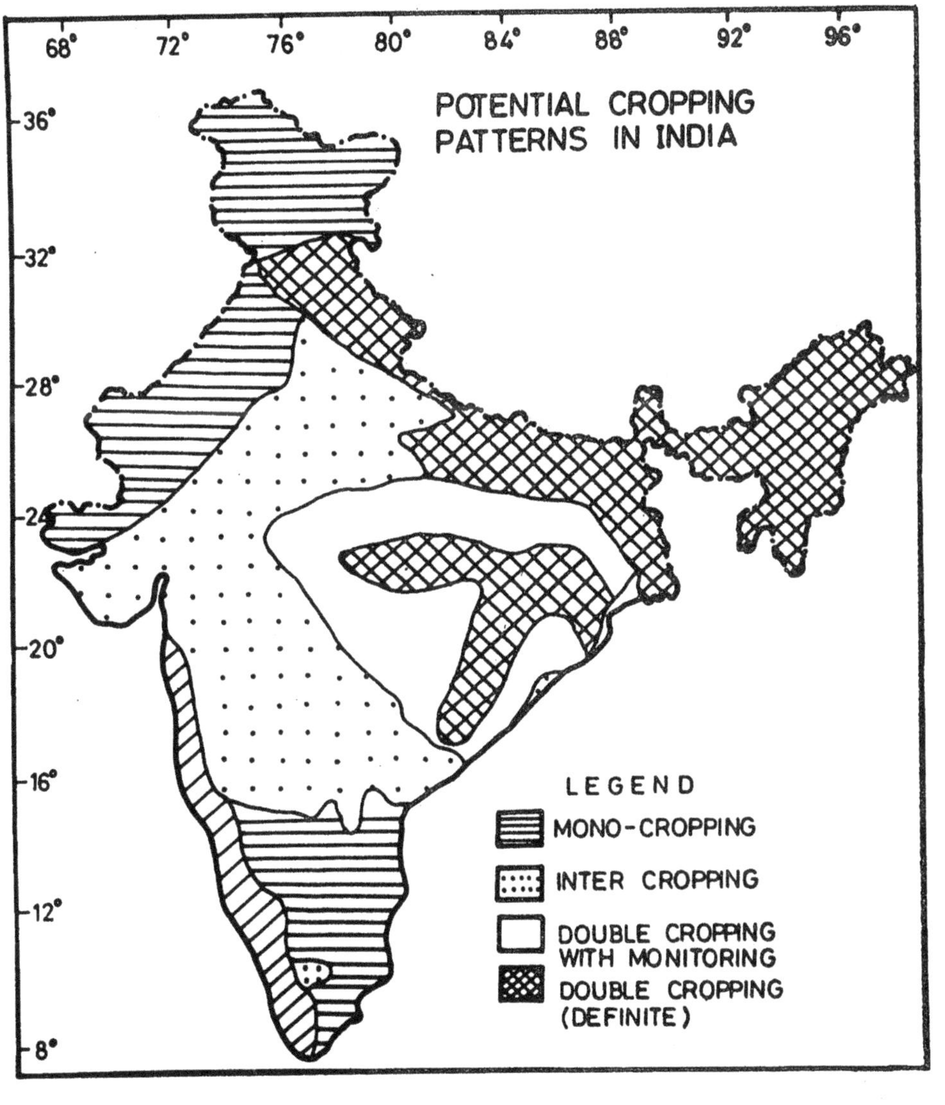

Fig. 2.4: Potential Cropping Patterns in India

conserved rainfall) is taken during a year. But with research based technology it is possible to raise the cropping intensity. The possible cropping intensity for different regions is given in Table 2.4. Taking an overview of rainfall, water holding capacity of soils and potential evapotranspiration, India's potential cropping system could broadly be shown in Figure 2.4. These classifications are of great help in planning proper agriculture strategies.

Table 2.4: Possible cropping intensity in different regions

Sr. No.	*Rainfall (mm)*	*Broad soil group*	*Growing season (week)*	*Cropping Season*
1.	350-650	Shallow Red and Black soils	20	*Kharif* (single crop)
2.	350-650	Sierozems and Alluvium	20	*Kharif* or *Rabi* on conserved rainfall
3.	350-650	Deep Black	20	*Rabi* (single crop)
4.	650-800	Alluvium, Black and Red	20-30	Intercropping in *Kharif*
5.	> 800	Alluvium, Black Red and Submontane	30	Double cropping (*kharif+rabi*)

Kharif - summer, *Rabi* - winter

Hydrological Cycle

Hydrology is the study of water cycle in nature. The ever-recurring conversion of ocean water to atmospheric water, precipitation, ground water. run off to ocean water with its many ramifications is called hydrological cycle or water cycle. The study of hydrology helps to understand the amount of water available to the crops during the various parts of the year and the effects of water on soil and plant. The hydrological cycle can be illustrated as under:

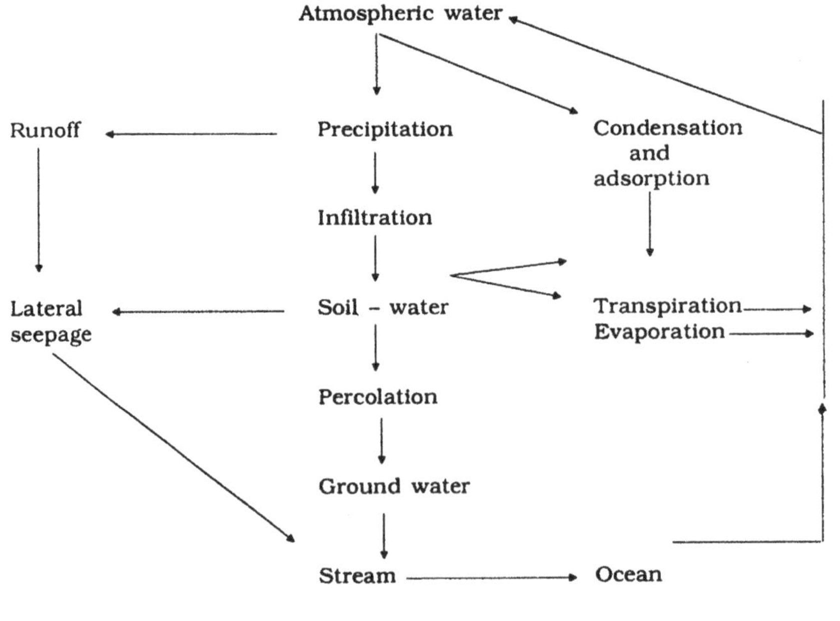

In order to determine the magnitude of the individual component of the hydrological cycle, measurements on these components can be made directly by using rain-gauge, run off stage recorder, weighing lysimeter for evapotranspiration and direct soil moisture determination by gravimetric method.

Rainfall: Rain may be classified according to amount (mm), intensity (mm/h) or the energy of impact (kg/m/ha). Rainfall is measured by an ordinary rain-gauge (Fig. 2.5) while a recording gauge (Fig. 2.6) gives information about the rainfall intensity. The intensity can be classified in to gentle, medium, heavy or severe as reported in Table 2.5.

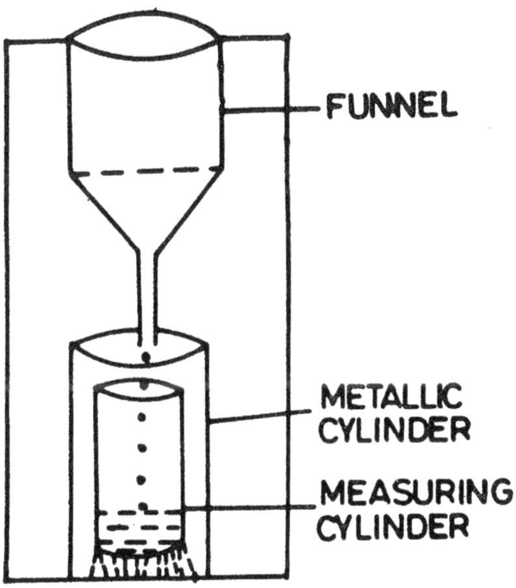

Fig. 2.5: An Ordinary Rain Gauge.

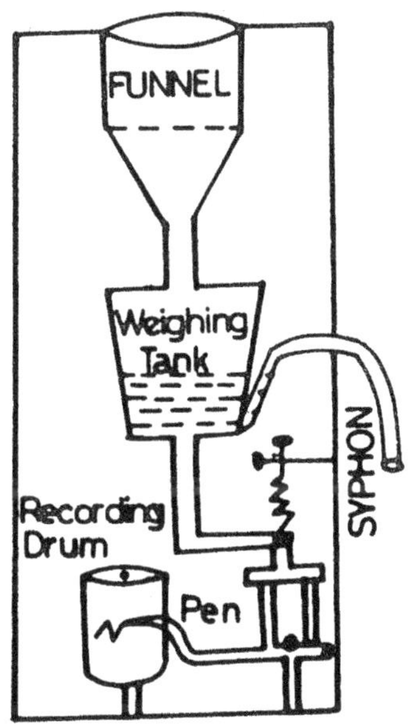

Fig. 2.6: An Automatic Rain Gauge.

Table 2.5: Classification of rainfall intensity

Rainfall rate (mn/h)	*Class*
< 6	Gentle
6–12	Medium
12–50	Heavy
> 50	Severe

Infiltration: It is the entering of water into the soil. Percolation is the downward movement of water through the soil. It is same as drainage. Interflow is lateral seepage of water in relatively pervious soil above a less pervious soil. Infiltration rate is the rate at which water enters into the soil and is measured by a double-ring-infiltrometer. Two metallic rings of diameter 30 and 60 cm respectively and height 20 cm are half embedded concentrically in the field by careful hammering without disturbing the soil. To measure infiltration rate of the soil, water is added first to the inner ring and immediately after to the outer ring such that the level of water (about 5-7 cm) in both the rings is maintained at a constant level. Immediately after adding water in the inner ring, the water level in the ring is measured with a scale and the loss in the head of water is recorded with time. A graph between infiltration rate and time is plotted to know the infiltration capacity of the soil which is the maximum rate of infiltration. It takes about 1.5 to 4.0 hours time to attain this value of infiltration. The various classes of infiltration are reported in Table 2.6.

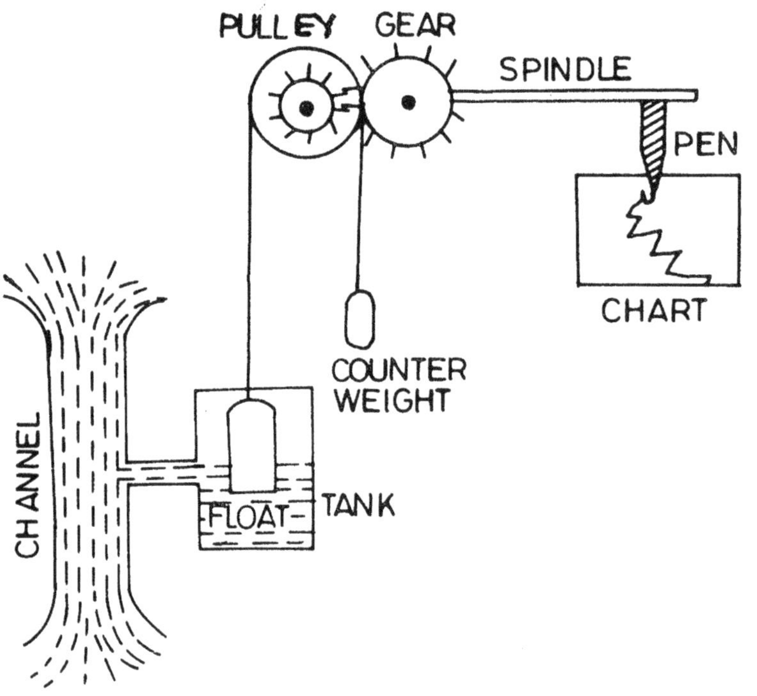

Fig. 2.7: A water stage recorder.

Table 2.6: Classification of infiltration capacity

Class	*Infiltration capacity, mm/h*
Very rapid	> 254
Rapid	127–254
Moderately rapid	63–127
Moderate	20–63
Moderately slow	5–20
Slow	1–5
Very slow	< 1

Surface Runoff: Runoff from a watershed occurs when the rate of infiltration capacity of the soil is less than the rainfall rate. It can be measured by suitable gauging of the watershed. The runoff is allowed to pass through a permanent channel located at the lowest point of the drainage channel. The channel is transversally connected with a water level recording tank. The level of water in the tank activates a pen on a float which leaves its mark on a clock driven calibrated chart (Fig 2.7). The runoff rate and its amount is known from the chart.

Ground Water: The infiltrating, percolating and seepage waters ultimately finds their way to the ground water after saturating the soil. The ground water level can be measured with a suitable scale.

Evapotranspiration: Evaporation from the land surface and transpiration from the plant surface jointly called evapo-transpiration which constitutes a major component of hydrological cycle. The evaporation of water from the land surface depends on the magnitude of the vapour pressure gradient from the soil surface to the atmosphere just above the ground and wind velocity near the ground. The vapour pressure of the atmosphere depends on atmospheric temperature and relative humidity while that of the soil surface depends on soil moisture content and temperature at the soil surface besides soil moisture supply below the soil surface. The soil temperature depends upon sunshine, soil colour, mulch and plant cover at the land surface. Transpiration in a given atmosphere depends upon plant type, its canopy and stage of growth and availability of soil-water to plant roots. Usually evapotranspiration (ET) is calculated by knowing other components of the following soils-water balance equation:

$$P + I = R \pm M \pm D + ET$$

where P is rainfall, I is irrigation, R is runoff, M is change in soil water content in the rhizosphere, D is deep percolation or capillary rise of water across the bottom of root zone. All measurements are recorded in the same unit (mm). The runoff is usually checked by providing effective bunds in the field.

Chapter — 3

Watershed Planning

The complex inter-relationships between the natural resources and its living population in a watershed must be explored with great precision while planning a watershed. A two pronged strategy—a short term to provide immediate benefit to the inhabitants and a long term approach for a full assessment of the problem and more practical and permanent solution to realize the full potential of the natural resources of the region, must be adopted right from planning of the watershed development programme. In the process the inhabitants of the watershed should become a partner not only because an imposed solution of a problem would be an act of thoughtless ignorance, but also because the solution would not succeed without essential local participation.

In a watershed, water is the most limiting component not only for agriculture but also for drinking and is not available through canals or ground harvesting. The planning there is proper conservation and storage of rainfall-runoff and its use for irrigating and other useful purposes which otherwise goes waste and causes floods in the downstream area. This idea suits to the people of the watershed and inspires them with unlimited enthusiasm and are always willing to do anything to get their land irrigated for reaping good harvest of crops. Once the interests of the people are economically linked with the conditions in the watershed, they will automatically appreciate the situation and will protect the watershed without any persuasion. They will participate emotionally if there economic security is assured and will then take care of the ecological security. Also they will participate physically with their labour and other inputs if they are assured that the project authority means business for their benefits.

Site Selection

A watershed is a geo-hydrological unit of an area that drains at a common point. The area to be selected should have acute shortage of water, large population of schedule caste and schedule tribes, preponderance of waste and common lands, contiguous with another watershed (if available) and actual wages are lower than the maximum wages prescribed by the government. Usually the funding agency provides guidelines for selecting the districts and blocks where such programmes are to be taken up. The area should have good potential for dryfarming and community of the area should be willing and cooperative to participate in the over all development activities for the area.

The site selection is the most technical job and must be done by a multi-disciplinery experts team. The ideal site should have natural advantage location, preferably around the foot hills, were there is a problem of flash floods and soil water conservation. The activities could later be extended to other areas including flat lands. While selecting the site the following major points should be kept in view:

1. The soil of the catchment should be heavy to permit high runoff even with protective vegetation cover.
2. The soil near the embankment should be fairly heavy to minimise seepage loss of stored water in the dam.
3. The dam site should be narrow and of both sides it should be stable.
4. There should be sufficient space for water collection in the dam.
5. The land to be irrigated should be nearby.
6. The site must have stable and natural spillway for safe disposal of excess rain water.
7. A good amount of rainfall goes as runoff and creates flood in the downstream land.

Watershed Delineation

The delineation of the watershed should be done by an expert team with the help of land topographic map of the area after completing its reconaissance survey. If the land contour map of the site is available from the Survey of India, the watershed can be identified easily from the contour lines. Without such a map, the watershed can be identified by visual experts observations of natural drainage lines discharging to a common stream from both the sides. The levels should be measured in different directions till one comes across the heighest level of points of evaluation. If these points are connected they will form ridge line of the watershed.

Agricultural watershed could be of size 3000 to 5000 ha. However, the normally recommended size for Indian situations is about 1000-2000 ha.The size is important from the completion of the development programmes within a limited period of time which is further governed by the available funds and technically trained personnel. The size of watershed would also differ with different stages between the micro and macro level planning and implementation of development programmes. Watersheds of smaller size have distinct advantage of involving a small number of village (1-2) within a resource limit that shares the common historical, social and economic pattern. It is therefore easier to get co-operation of people for group action in soil and water conservation programme as well as its management options. According to recent guidelines from the Ministry of Rural Development, Government of India, each watershed could be about 500 ha covering atleast one village.

Formation of Various Committees

For proper planning and execution of watershed development programme, the following committees must be constituted.

Project Implementation Agency (PIA)

Such programmes are administered by the district authority (Zilla Parishad, DRDA), but it is desirable and necessary to involve voluntary agencies and other institutions such as Universities, Agricultural Research and Training Institutes, Corporations, Cooperatives, Banks, Public and Private Commercial Departments in planning, coordinating and supervising the formulation and implementation of watershed development projects in groups of selected villages. The role of PIA will be to motivate the Gram Panchayats to pass the necessary resolutions to make public contributions, conduct participatory rural appraisal (PRA) exercises, to prepare the development plans for each watershed, undertake community organization and training for the village communities, provide technical guidance and supervision of watershed development activities, manage project implementation, inspect and authenticate project accounts, undertake action research to adopt low cost technologies and/or validate and build upon indigenous technical knowledge, monitor and review the overall project implementation and set up institutional arrangement for post project operation and maintenance and further development of the assets created during the period. Also state government departments such as Agriculture, Forestry, Horticulture, Animal Husbandry etc. or other government insti- tutions may act as Project implementation agencies.

State Watershed Programme Implementation and Review Committee (SWPIRC)

To ensure coordination amongst various Government Departments, Agricultural Universities, Voluntary Agencies (VAs) and Training Institutions, a state level Watershed Development Implementation and Review Committee shall be constituted under the Chairmanship of the Chief Secretary/Additional Chief Secretary/Agricultural Production Commissioner/Development Commissioner/ Secretaries/Heads of Departments of related departments, Director of Research of State Agricultural University/Directors of a few state level training institutes like State Institute of Rural Development (SIRD)/Institute of Administration/Management, five or six representatives of important VAs in the state who are involved in watershed development programme. This committee will meet atleast twice a year to monitor, review and evaluate the progress of implementation of the watershed development programme, particularly with reference to involvement of VAs and for funding arrangements from different programmes. The committee would also have the power to relate the cost norms in specially difficult areas and the area ceiling required for taking up a watershed. This committee will be constituted by Development Secretary.

Watershed Development Advisory Committee (WDAC)

The Zilla Parishad (ZP/District Rural Development Agency, DRDA) may constitute a WDAC under the chairmanship of the Chief Executive Officer of the Zilla Parishad consisting of 3 or 4 members from amongst the multi- disciplinary team in the Zilla Parishad/DRDA, 5-6 representatives of VAs/PIAs from the district and 1-2 members from the relevant Research and Training Institute located in the district. This Committee will advise PIA on various aspects of watershed development.

Where the Voluntary Agencies (VAs) or other institutions are not available to work as PIA, the ZP/DRDA may act as PIA in the same manner as other PIAs. However, ZPs/DRDAs may ensure that no permanent staff is employed for this programme.

Watershed Development Teams (WDT)

Each PIA shall carry out its duties through a multi-disciplinary team designated as the Watershed Development Team (WDT). Each WDT may have alteast four members—one each from the disciplines of plant sciences, animal sciences, agricultural engineering and social sciences or post graduation in botany, economics, sociology/social work. Those who have practical experience in the rural areas should be given preference. The senior most among them shall be designated as the project leader. The WDT shall work exclusively and full-time for the watershed dvelopment projects in the selected 10-12 villages. The PIA will be at liberty to either earmark its own staff exclusively for this work or recruit fresh candidates including retired personnel or take people on deputation from government or other organization. The establishment charges for the WDT shall be provided within the limit prescribed by DRDA/ZP. A PIA may take up several groups of villages by constituting the required number of WDT's. The WDT shall be located at the PIA/block or local head quarter/any other small town whichever is the nearest to the cluster of slected villages.

Watershed Association (WA)

A Watershed Association of all the beneficiaries should be formed and registered as a society under the Registration of Societies Act. This society should be headed by an elected Chairman—a village sarpanch or any other influencial man of the watershed area and a Secretary. This association will meet, atleast, twice a year to evolve/improve the watershed development plan, monitor and review its progress, approve the statement of accounts, formation of user groups/self help groups, resolve differences/disputes between different user groups and collect donations/contributions for the watershed and laydown procedures for the operation and maintenance of assets created, approve the activity that can be taken up with the money available and other related matters. The Secretary of WA shall assist the Chairman in discharging the responsibilities entrusted to the WA and will be a full time paid employee.

The Watershed Committee

The day to day activities of the watershed development projects shall be carried out by a Watershed Committee comprising of about 10-12 members belonging to user groups (4-5), self-help groups (3-4), gram-panchayat (2-3) and a member of the watershed development team giving adequate representation of women SC/ST/shepherd community etc. This team will elect a chairperson from amongst themselves and will work under guidance and supervision of WA. This committee will be responsible for co-ordination and liaisoning with the gram panchayat, WDT, DPAP/ZP and other agencies concerned to ensure smooth implementation of WDP.

After constitution of these committees, the watershed development will start functioning for undertaking various development activities as per the priorities listed in the master plan prepared for this purpose.

Requirements for Watershed Planning

The following requirements must be met for planning watershed:

i) Reconnaissance survey of the watershed area for delineating nature of soil-water conservation and its related problems, their probable remedial measures and requirements for its overall development. It should be followed by preliminary and detailed surveys for basic resources. Land area map should be prepared on suitable scales for carrying out these surveys.
ii) Preparation of soil map and classification of land for uses according to capability for agriculture, horticulture, forestry, pasture development, rearing of cattle/sheep etc.
iii) Preparation of inventory of existing land uses, farm sizes and fragmentations.
iv) Appraisal of agricultural production pattern and potentials.
v) Topographic features and hydrology of the watershed area,
vi) Geophysical and geo-hydrological surveys to delineate areas suitable for ground water development.
vii) Information on rainfall intensity, amount and pattern, on weekly and daily basis.
viii) Formulation of an integrated time based plan.
ix) Fixation of priorities for the implementation of the programme.
x) Assessment of benefit cost ratio.

xi) Social and political forces and their involvement in the wathershed development programme. In many watersheds, economical and political powers reside at the bottom of the watershed where there is a valley and therefore are not influenced much by soil conservation problems. Rather they are benefited from the run off. Usually they get maximum benefit of the watershed development which is against the policy of equal benefits to all.

Basic Resource Surveys

The first and foremost step for efficient planning of watershed is the basic resource surveys of the watershed area which include:

i) ***Soil and Climatic Survey***: This survey should be carried out by the Project Implementation Agency, in consultation with the scientists of the research organization located in the region. The information to be generated includes, soils origin, mineralogy, texture, structure, infiltrability, water holding capacity, depth, concretion layer depth, erodibility, pH, EC, sodicity, organic matter, nutrients status etc. The climatic data should include rainfall amount, intensity (daily and weekly basis), distribution, daily evaporation, potential evapotranspiration, relative humidity, maximum and minimum temperature, wind velocity, sun-shine hours, besides frequency and extent of frost/hailstorm, if any. Based on weekly rainfall (at a certain confidence level, usually > 50 per cent) and potential evapotranspiration and expected soil stored water at the planning time (in 1 metre depth), weekly water availability period (Appendix 1), and accordingly based on available water (both raintall and irrigation, if available), potential crops and cropping systems are identified. Also the extent of various soils related problems such as soil crusting, presence of clay pan/calcium carbonate concretion layer at shallow depths, salinity, sodicity, nutrients deficiency for the potential crops are identified and their remedial measures are suggested. For soil survey the whole watershed area should be measured and divided into suitable grid pattern (100 × 100 m or 80) and the soil sampling is done in each grid and a soil map is prepared on a suitable scale (1:50,000). The various types and requirements of different types of soil surveys are given below.

Types of Soil Surveys

The following types of soil surveys are commonly conducted :

a) ***Reconnaissance Survey***: The project executive gathers general information about types of soils and its suitability for different purposes. The mapping is less elaborate and small scale base maps are used. The field traverses are made at convenient intervals and the survey is carried out in a short time.

b) ***Detailed Survey***: The detailed surveys are conducted to provide maximum information required for land use classification. The mapping of the land is done in an elaborate way depicting the lowest category of mapping units. Large scale base maps used for showing detailed information and field observations on soil properties are made at short interval in the field. This survey is done for agricultural land suitable for crop production.

c) ***Detailed Reconnaissance Survey***: These surveys are done in combination of both reconnaissance and detailed surveys. The area of land with low production potentials are surveyed according to reconnaissance survey and good lands are surveyed in depth.

A brief account of requirements of different surveys is given below:

Parameters	*Reconnaissance survey*	*Detailed survey*	*Detailed reconnaissance survey*
Base maps and scale	Small scale Base maps Topo sheets	Large scale Base maps	Cadastral maps. Areal photograph (1:25.000)
Averages	Areal photographs 1:50,000	Cadastral maps 1:5,000	Partly for reconnaissance and partly for detailed
Mapping units	Soil series associations	Series, type, phase	-do-
Observations interval (Distance of Auger hole)	Wide interval (750 m)	Close interval (250 m)	
Profile examinations	3–6 km	2–3 profile per 240 ha	-do-
Soil boundary observations and demarcation	At intervals	Entire length	-do-
Scale of publication	1:1 to 1:5 lac	1:5,000 to 1:25,000	-do-

Soil Survey Equipments

i) Soil augers—post hole, tube screw and powered.
ii) Implements—spade, pick-axe, crow bar, khurpi, knife, basket.
iii) Base—maps, munsel colour chart.
iv) Dilute hydrochloric acid, magnifying glass, Abney's level, compass, measuring tape, sample bags, labels and water bottle.

Soil Survey by Traverse Method

It consists of examination, classification and mapping of soils in the field by using above mentioned equipments in required interval of area usually on grid basis. Soil boundaries are checked by examination in the soil profile (Appendix 2) in the field at designed grid interval. The soil properties commonly recorded for soil classification purposes are soil depth, texture, permeability of surface and sub-soil, land slope, erosion hazard, soil colour, salinity, fertility, water holding capacity, stoniness, undulation, EC, pH, available plant nutrients, and mechanical fractions-sand, silt and clay. Classes of these properties required for land classification are reported in Table 3.1.

For soil survey studies soil series and soil type are considered relevant unit for soil classification. It is usually given the name and the place or geographic area where the soil profiles are first examined and recognized such as *Ladwa* series. It is named by adding texture (Appendix 2) class name to a series name such as Ladwa silty loam.

Table 3.1.: Classes of soil erosion, depth, permeability and slope

Soil erosion classes

Symbol	*Erosion class*	*Characteristics*
e_1	Slight	0 – 25% top soil removed
e_2	Moderate (sheet & rill)	25 – 75% top soil removed
e_3	Severe	75 – 100% top soil removed
e_4	Very severe	Gullied land
e_5	Very very severe	Deep gullied land or sand dunes

Soil depth classes

Symbol	*Name*	*Depth range (cm)*
d_1	Very shallow	0 – 7.5
d_2	Shallow	7.5 – 22.5
d_3	Moderately deep	22.5 – 45.0
d_4	Deep	45.0 – 90.0
d_5	Very deep	> 90.0

Permeability classes

Symbol	*Permeability class*	*Rate of flow (cm/h)*
1.	Very slow	< 0.13
2.	Slow	0.13 – 0.50
3.	Moderately slow	0.50 – 2.00
4.	Moderate	2.0 – 5.00
5.	Moderately rapid	5.00 – 13.00
6.	Rapid	13.00 – 25.00
7.	Very rapid	> 25.00

Slope classes

Symbol	*Slope class*	*Slope %*
A	Nearly level	0 – 1
B	Gently sloping	1 – 3
C	Moderate sloping	3 – 5
D	Strongly sloping	5 – 10
E	Moderate Steep	10 – 15
F	Steep	15 – 25
G	Very steep	25 – 33
H	Very very steep	> 33

Symbolic Description of Soils

The soil series could be described symbolically as

$$\text{Ladwa}\ \frac{d_3 - C - 0.8}{A - e_2}$$

The above symbol indicates that the soils are of Ladwa series which is moderate deep (d_3), clayey (C) in texture, with low permeability (0.8) of soil profile. The lands are nearly level (A or slope 1%) and there is only slight erosion (e_2).

Effect of Climate

Climate has a great influence on the capability of the land to produce agriculture, pasture or forest crops. The land capability classification decreases with the decrease of the effective rainfall. The ill effect of low rainfall can be reduced by providing irrigation to the land. Based on the climate, the land can be classified as reported in Table 3.2.

Table 3.2. Land capability classification based on climate.

Type of climate	*Land capability class*
(i) Humid with well distributed rainfall	I
(ii) Humid with dry spells	II
(iii) Sub humid	II-III
(iv) Semi arid	III-IV
(v) Arid	V

(ii) Land and Engineering Survey

This survey should be done by a team of land surveyers and includes parameters like land area, length-width ratio, elevation, depth, shape, physiography, vegetation (temporary/permanent) nature and extent of slopes, runoff characteristics, waterways—their length, size and depth, number and extent of gullies, ravines and depressions for storing water and final draining point for the runoff besides the depth of underground water table and its quality. Based on these surveys, land contour maps of the watershed area and water table depths should be prepared and location and extent of engineering works like soil and water conservation, water retention, dams/ponds, gully plugging, spillway, culverts, percolation tanks, diversion bunds, grade stabilization structures, bunding (contour/graded) should be identified and properly designed and cost estimates are prepared. The types of existing vegetation and their extent of coverage and development of land for forest, grasses, horticulture, crops and animal grazing purposes should be determined on the basis of land capability classifications. Properly ownership and road maps should also be procured from the concerned government organization.

Land Capability Classification

The grouping of soil into different classes, sub-classes and units is done primarily on the basis of their ability to produce plants without deterioration for a long time and is based on the effect of the combination of climate and permanent soil characteristics on the risk of soil damages, limitation in use, productivity and management requirement. The land capability class is further divided into sub-classes depending upon the kind of limitation for its use. These sub classes are climate (c), erosion (e), soil (s) and wetness (w). Their details are as follows:

(i) c (climate): Limitation of rainfall and temperature.

(ii) e (erosion): Risk of erosion or degree of damage that has already occurred due to erosion.

(iii) s (soil hazard): Limitations of root zone, water holding capacity, drainage, salinity, sodicity, stoniness etc. which may affect tillage operation, rooting depth of crops etc.

(iv) w (wetness): Limitations of wetness, drainage or overflow.

In case there are two kinds of limitations, both can be indicated, the dominant one is shown first. If both the limitations are equal, the symbols for sub-classes are used in order of priority i.e. e, w, s and c.

Based on soil and land survey the land could be grouped into following eight classes.

Class I: Levelled, deep and productive lands fit for intensive cropping with no special management practices. Such lands have a high water holding capacity, good internal drainage, medium to rich in plant nutrients.

Class II: Gently slope (< 2% slope), deep and productive land, good for cultivation but having a minor problem of soil erosion, drainage salinity or alkalinity and needs some special management practices for regular cultivation.

Class III: Slope (2-5%), medium deep lands which could be put under cultivation with some special soil conservation and other management practices. Such lands will have erosion, salinity and alkalinity problems.

Class IV: Productive lands with land slopes of 5-8 per cent and having major erosion problems which require special measures for their control.

Class V: Good productive mountain land (slope 8-12%) prune to gully erosion problem and shallow soil depth, suitable for pasture development.

Class VI: Steep slopy (slope >12%) lands suitable for pasture, forest, etc.

Class VII: Very steep slopy, highly eroded land suitable for pasture and forest.

Class VIII: Highly eroded gullied or sand dunal lands suitable for wildlife conservation project.

Classes I to IV are considered suitable for cultivation and providing irrigation facilities. Class V lands have several limitations for cultivation while Class VI lands are not considered fit for cultivation and irrigation. Classes VII and VIII lands are suitable for pasture/forest and wildlife conservation only as they are unfit for cultivation.

Land Capability Maps, Coding and Colour Notation in Classification

Land capability maps are prepared by interpreting the soil survey data. Different land capability classes, sub classes are differentiated on the basis of major soil characteristics and land feature. Each land capability class and subclass are coded and given standard colour (Table 3.3) after demarking the area of the class.

Table 3.3. Colour notation for different land capability classes.

Capability class	*Colour notation*
I	Light green
II	Yellow
III	Red
IV	Blue
V	Dark green
VI	Orange
VII	Brown
VIII	Purple

Coding is usually done with the help of the soil survey data collected during soil survey after locating the site on the map. Out of all the soil characteristics and associated land features, the most important ones considered for classification are effective depth, texture of the top soil, permeability of top soil and that of sub-soil, slope and erosion hazard. The first four are written in the numerator and the remaining two in the denominator along with the name of the soil series.

$$\text{Series name} - \frac{\text{Depth} \times \begin{matrix}\text{Texture of}\\ \text{the top soil}\end{matrix} \times \begin{matrix}\text{Permeability of}\\ \text{the top soil}\end{matrix} \times \begin{matrix}\text{Permeability}\\ \text{of sub–soil}\end{matrix}}{\text{Slope} \times \text{Erosion}}$$

For example Class III – MC Ladwa – $\frac{d_3 \times L \times 3 \times 3}{A-e_2}$

It shows that the soils are Ladwa series, moderately deep, loam in texture, moderately slow permeable, nearly level land with moderate erosion. The land could be put to class III MC as the texture of soil is loam (M, medium texture) and Ladwa is situated in semi-arid climate (c).

(iii) Socio Economic Survey

A watershed is the natural unit for economic management and therefore its survey must involve view point of individual farmer, farming community, individual person and execution department. The community have their needs, custom values and economic status. Therefore the various measures taken for the treatment of watershed should match with the needs of people. Proper management of watershed, therefore, requires not only the knowledge of application of scientific techniques but also substantial investment of capital and labour whose returns may be high only in the long run and thus incompatible with the prospects of the farmer. How to motivate the farmers to make necessary investment and sacrifice is a socio-economic problem. The most important economic aspect of the watershed management is to determine economic feasibility of the investment on the development programmes. There are alternate techno- logies of soil conservation, water harvesting, storage and conveyance of water, alternate cropping patterns and technologies of water conservation and package for making optimum use of available water and other resources. The job of economist is to evaluate these alternatives and suggest the efficient one. Several techniques like benefit-cost analysis, interval rate of return etc. are available for evaluating such projects.

It is not necessary that the cultivation of crops is always advantageous. It could be agro-horticulture, agroforestry/silvi pastoral/livestock rearing or combination of several of these aspects which is more remunerative. In the hills livestock is more promising than crop production. One of the strategies of watershed management in the hills is the encouragement of social forestry, forage husbandry and improvement of cattle. It is necessary to provide feed for cattle and find ways and means so that it does not cause any damage to forest. Unless and until people's need for fuel, fodder and timber is not met from other alternative sources, damage to forest is bound to occur. For rolling ground, agroforesty/agro-horticulture may be more promising while for flat lands crop cultivation could be of high value. For economy feasibility study the following aspects need to be considered:

i) Daily need for social aspects of the village community.
ii) Improvement of ecological balance and preservation of natural resources and

iii) Development of backward areas and rural upliftment, employment and ancillary industries.

Social and legal status information can be obtained from census report, revenue records, district gazetteers, legislation etc. Such records are required for determining sufficiency/deficiency in food and other agricultural stuff-fibre, fodder, fuel, animals and man-power, level of acceptance/adoption/follow up of the programme by the community and their basic requirements, operational amenities and to determine responsive and priority areas of development. Information which are not readily available should be collected by conducting socio-economic survey of the village (Appendix 3). Through this survey information with regards to land holding, sources of income of the village community, educational qualification of farmers family members, size of family, existence of infrastructural facilities like, school, hospital, bank, development organization etc. is gathered.

People Participation

In planning watershed development programmes, community action is indispensable. Small farms, dominating in many developing countries, do not allow for an individual farmer to have an impact on land improvement. Watershed approach integrate all farms and farmers of the area for development of the land and other resources. Soil conservation programmes are tackled on a watershed basis, as a concerted action, where all efforts reinforce and complement each other. Some erosion control measures may be within the reach of a group of farmers but for larger undertakings the support of Government is necessary. Infact, the Government is required to maintain the basic natural resources—the land and water. Such programmes should give priority to continuous growing of agriculture rather than on economic return measures. At all times, farmers should be closely associated in the planning and execution of field programmes. New technology should not become a burden to the farmers. Rather it should accrue benefits to the farmers.

The need for public participation is especially important in watershed schemes when the project treatments involve:

(a) private or common land;
(b) temporary land closures,
(c) changes in land management practices (e.g.; grazing, cut and carry, rotational management, etc.);
(d) investment of time and/or financial resources by local inhabitants (e.g.; planting vetiver grass hedges or private construction of water harvesting structure).

Several approaches to soliciting popular participation as well as monitoring beneficiary impact of watershed investments have been commonly used in the field. These approaches include:

(a) interaction with local people (personal interview, questionnaires, household surveys, individuals observations);
(b) communication and interaction with community leaders;
(c) dissemination of information and participation through community meetings; and.
(d) interaction through community organizations (non-governmental organizations; NGOs, religious organizations, local councils, regional organizations).

The operational challenge for watershed development practitioners is two-fold: (1) how to efficiently obtain information and opinions from the rural residents; and (2) how to convert and integrate that information into programme which are locally relevant, technically sound and which also serve broader.

In most developing countries, villagers are dependent on land to graze animals, to take timber for making agricultural implements and to cut fuel wood. The problem is so real that it makes difficult to manage even the reserved forest. Time scale fragmentation of land due to family division also creates, problems in managing a developed watershed.

Preparation of Master Plan

The various programmes to be taken up for the development of the watershed area must be documented in the form of a master plan (Appendix 4). The master plan should include:

(i) Introduction of the watershed area
(ii) Characteristics of the watershed—general information, climate, soils, natural vegetation, present land use and capability classification, proposed improvement in land, animal, human and other resource use and socio-economic improvement.
(iii) Analysis of problems and potentials of watershed.
(iv) Improved recommendations for crop management, soil and water management, pasture, forestry and horticultural plantation.
(v) Schedule of development activities and operational programmes.
(vi) Cost-benefit analysis and budget estimates.
(vii) Supplemental informations—various committees members, etc.
(viii) Monitoring and evaluation.

It should also include various maps—index map, contour map etc., and schedule for training of farmers and monitoring and evaluation mechanism for the watershed. All villages/community must be associated from the initial planning and execution of the programme to the final stage of development and its subsequent maintenance and management. The resources benefits must be distributed by the community among its people.

The execution of the entire watershed programme must be strictly taken up as per the master plan after full involvement and support of the village community. Efficient management of dryland agriculture on watershed basis is the key to improve and stabilize agricultural production of dryland areas and to bring prosperity in the region. However, this break-through is possible with sincerity, honesty and dedication of various bodies involved in this programme.

Chapter — 4

Rainfall Runoff Estimation

Estimation of total amount of runoff from a catchment is required for assessing the size of farm pond/reservoir and that of peak rate of runoff for designing the various engineering structures (culverts, spillways etc.) for flood and soil erosion control and for the safety of the reservoir. These estimates are not simple. Literally thousands of papers have been published reporting the results of data processed according to various theoretical approaches. These vary from simplistic models to highly complex mathematical models based on a large number of watersheds parameters and are solvable only with the use of computer. However, basic data required for operating various mathematical models are usually not available in most of the developing countries as the streams are not gauged for these observations. However, the estimation methods are used for knowing total runoff.

Estimation of Total Runoff

The following two methods are commonly used for the estimation of total runoff.

(i) Average runoff coefficient method and
(ii) The curve number method

The various details of these methods are reported as under:

Average Runoff Coefficient Method

The average runoff coefficient method is crude but is the simplest of all the methods and therefore is more commonly used. The total quantum of runoff is determined by multiplying the average rainfall of a period by a runoff coefficient (Table 4.1). Considerable judgement is required for selecting the runoff coefficient which varies with amount, intensity and distribution of rainfall, physiography and topography of the land, land slope, and soil type and its moisture status. It is therefore desirable to divide a catchment into a large number of micro-catchments such that each micro-catchment is hydrologically uniform. Then an appropriate runoff coefficient value for each of the micro-catchments is assigned using Table 4.1 and then weighted runoff coefficient (c) for the entire watershed is determined by the expression:

Table 4.1: Runoff coefficients for different soil textures, soil covers and land slopes

Vegetative cover and land slope	*Soil texture*		
	Sandy loam	*Clay and silty loam*	*Stiff clay/rock*
Wood land			
0-5% slope	0.10	0.30	0.40
5-10%	0.25	0.35	0.60
10-30%	0.30	0.50	0.60
Pasture land			
0-5%	0.10	0.30	0.40
5-10%	0.16	0.36	0.55
10-30%	0.22	0.42	0.60
Cultivated land			
0-5%	0.30	0.50	0.60
5-10%	0.40	0.60	0.70
10-30%	0.52	0.72	0.82

$$C = \frac{A_1 C_1 + A_2 C_2 + \ldots\ldots A_n C_n}{A_1 + A_2 + \ldots\ldots A_n} \qquad \ldots (4.1)$$

where C_1, C_2, ... C_n and A_1, A_2............. A_n are respectively runoff coefficient and land area for each of the micro-watershed.

The total runoff (Q) is then calculated by the relation.

$$Q = RCA \qquad \ldots (4.2)$$

where R = rainfall and A = area of the watershed.

The Curve Number Method

The United States Soil Conservation Services (USSCS) developed the following formula for estimating total runoff (Q in mm) based on different land, uses, hydrological soil covers and soil moisture conditions :

$$Q = \frac{25.4\,(R - 0.2S)^2}{(R + 0.8S)} \qquad \ldots (4.3)$$

where S = potential maximum surface detention and is worked out by the equation:

$$S = \frac{1000}{CN} - 10 \quad \text{... (4.4)}$$

where CN is curve number. The curve numbers for different land uses, hydrological soil covers and soil moisture conditions are given in Tables 4.2, 4.3, and 4.4.

Table 4.2: Runoff curve number for different land uses and hydrological soil-covers at medium soil moisture level

Land Use	*Hydrological Soil Group*			
	A	*B*	*C*	*D*
1. **Open Space**				
a) Grass cover 75% area	39	61	74	80
b) Grass cover 75-50% area	49	69	79	84
2. **Fallow: Straight Row**	77	86	91	94
3. **Row Crops**				
a) Straight row—Poor condition	72	81	88	91
—Good condition	67	78	85	89
b) Contoured row—Poor condition	70	79	84	88
—Good condition	65	75	82	86
c) Contoured row—Poor condition on terrace	66	74	80	82
—Good condition	62	71	78	81
4. **Small Grain Crops**				
a) Straight row—Poor condition	65	76	84	88
—Good condition	63	75	83	87
b) Contoured row—Poor condition	63	74	82	85
—Good condition	61	73	81	84
c) Contoured row—Poor condition on terrace	61	72	79	82
—Good condition	59	70	80	81
5. **Closed Seeded Legume Crops**				
a) Straight row—Poor condition	66	77	85	89
—Good condition	58	72	81	85
b) Contoured row—Poor condition	64	75	83	85
—Good condition	55	69	78	83
c) Contoured row—Poor condition on terrace	63	73	80	83
—Good condition	51	67	76	80
6. **Pasture**—Poor condition	68	79	86	89
—Fair condition	49	69	79	84
—Good condition	39	61	84	80
7. **Forest**—Poor condition	45	66	77	83
—Fair condition	36	60	73	79
—Good condition	25	55	70	77

For low and high soil moisture levels, Table 4.4 should be consulted for determining the corresponding curve number.

Table 4.3: Characteristics of different hydrological soil groups

Group	*Soil characteristics*	
	Texture	*Infiltration rate, mm/h*
A	Deep sand	7.5 - 12.5
B	Sandy loam	3.7 - 7.5
C	Shallow sandy loam/clay loam	3.7 - 1.2
D	Clay	< 1.2

Table 4.4: Curve number for different soil moisture conditions

Soil - moisture level

Low I	*Medium II*	*High III*
100	100	100
87	95	99
78	90	98
70	85	97
63	80	94
57	75	91
51	70	87
45	65	83
40	60	75
35	55	75
31	50	70
27	45	65
23	40	60
19	35	55
15	30	50
12	25	45
9	20	39
7	15	33
4	10	26
2	5	17
0	0	0

Estimation of Peak-rate of Runoff

The USSCS developed following expression to estimate the Peak-rate of runoff (maximum rainfall intensity), Q_p in $m^3\ s^{-1}$:

$$Q_p = 0.0208\ A.Q\ \sqrt{T_p} \qquad \text{... (4.5)}$$

where A is area of watershed in hectares, Q is total rainfall runoff (cm) and T_p is time to peak rate of runoff of the hydrograph for a given rainfall event and is given by

$$T_p = T_c^{1/2} + 0.6\ T_c \qquad \text{... (4.6)}$$

where T_c is the time of concentration in hours and is defined as the time in hours required for rain water to flow from the most remote point of the watershed area to the outlet of the watershed.

The peak rate of runoff can also be estimated by the rational formula given by Ramser

$$Q_p = \frac{CIA}{36} \qquad \text{... (4.7)}$$

where C is the coefficient of runoff, I is rainfall intensity (cm/h) for a duration equal to the time of concentration and for the designed frequency (say for 25 year return period) and A is watershed area in hectare. A modified form of rational formula for Ambala Siwaliks has been suggested (Erasmus and Bansal, 1961) as under:

$$Q_p = 0.778\ CIA^{0.73}/36 \qquad \text{... (4.8)}$$

The runoff coefficient is defined as the ratio of peak runoff rate to the rainfall intensity. The time of concentration (T_c) is calculated by using the following empirical formula:

$$T_c = 0.0195\ K^{0.77} \qquad \text{... (4.9)}$$

and $K = L^{3/2}\ H^{-1/2}$

where T_c is time of concentration (minute), L is maximum length of channel reach (metre) and H is difference in elevation (metre) between the most remote point and the outlet of the channel. This relationship gives good estimates for watersheds having area more than 5 Km^2. For smaller watershed the formula is

$$T_c = 0.0195K^{0.77} + \left[2L_0\sqrt{n}/S_0\right]^{0.467} \qquad \text{...(4.10)}$$

where L_o is length of overland flow (m), S_o is slope along the path of water flow and n is Manning type roughness coefficient (Table 4.5). The second term in the above equation accounts for the overland flow.

Table 4.5: Values of Z and Manning's 'n' for different soil and land conditions

Soil texture	*Z = H/V*	*Land condition*	*n*
Sandy and loam sand	3 : 1	Earth	0.022
Sandy loam	2 : 1	Rocky	0.040
Loam and silt loam	1 : 5 : 1	Weedy	0.035
Clay	1 : 1	Short grass	0.045
Stiff clay	0 : 5 : 1	Medium grass	0.055
Rock	0 : 25 : 1	Long grass	0.095

In watersheds the overland flow phase effectively has a maximum travel time because both laboratory and field studies have shown that overland flow conditions can seldom be maintained for a distance longer than 100-150 m. Thus, the value of this term has a maximum value regardless of watershed size because of the upper limits specified for L_O. However, as the size of watershed increased the maximum length of channel (L_O) flow tends to increase. This means that, as watershed size becomes large, the channel flow component of T_c (equn. 4.10) will dominate the overland flow. The above equation is therefore expected to be valid for all sizes of watersheds.

The time of concentration can also be assessed from estimates of the velocities of overland flow and the channel flow. If the overland flow traverses more than one kind of surface, the travel time for each of the surfaces should be added up.

Peak Rainfall Intensity for a Certain Frequency

For calculating the expected peak rainfall intensity, the available historical data of maximum rainfall intensity recorded at the nearest rain guage station are arranged in the descending order of intensity. If N is the number of years for which rainfall records are available, T is the frequency (recurrence interval i.e. period of return) and n is the ranking of the severe storm, then N is given by

$$N = T \cdot n \quad \text{... (4.11)}$$

For example, peak rainfall intensity records are available for 20 years (Table 4.7) for a particular place and it is designed to select the storm (rainfall intensity) that will occur once in 10 years, then from the equation (4.11), we have n equal to 2. When all rainfall intensities (peak) are arranged in descending order, the second highest intensity in that order is the peak rainfall intensity which is expected to occur once in 10 years. The usual return periods which are selected for earthen dams, for water storage tanks, small dams for gully control, temporary check dams and other engineering structures are 25, 15, 10 and 8 years, respectively. At several places in India as well as in other countries rainfall intensity records are not available. The peak rainfall intensity (I, cm h^{-1}) in such a situation (for India) can be estimated by the relationship:

$$I = k \cdot T^a (t + b)^n \quad \text{... (4.12)}$$

where a, b, n and k are constants depending on the rainfall characteristics of the area, T is the return period and t is the duration of storm. For different zones of India, the values of a, b, n and k are given below in Table 4.6.

Table 4.6: Values of a, b, n and k for different zones in India

Zones	*a*	*b*	*n*	*k*
Northern	0.1623	0.50	1.0127	5.9160
Central	0.1712	0.75	0.9699	7.4645
Eastern	0.1353	0.50	0.8801	6.9330
Southern	0.1523	0.50	0.9465	6.3110
Western	0.1645	0.15	0.7327	3.9740

Tejwani *et al.* (1975) analysed the rainfall data of different rainfall stations in India and gave 24 hours rainfall intensity for different frequencies (Fig. 4.1). These values of I can be converted to rainfall intensity for any time I_t (usually the time that corresponds to the time of concentration) by the relation

$$T_c = I \frac{T + I}{t + I} \qquad \text{... (4.13)}$$

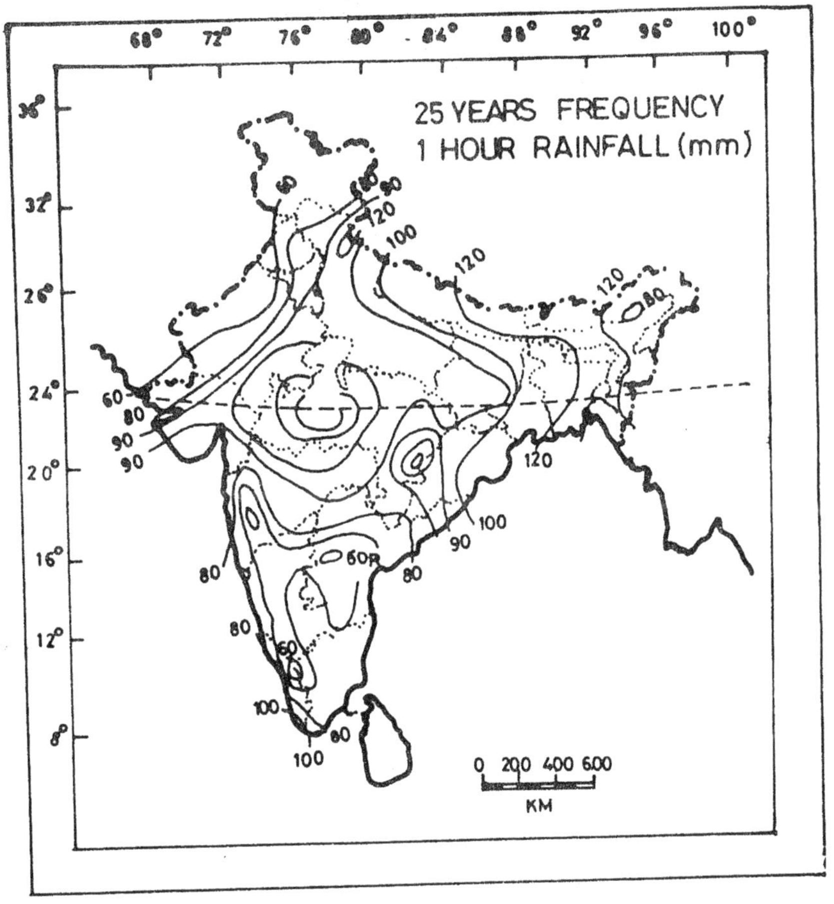

Fig. 4.1. Rainfall Intensities at 25 Years Frequency in India

where T is 24 hours intensity and t is the time of concentration. It can also be estimated from the graph between one hour rainfall intensity and rainfall intensity for any duration (Fig 4.2). The relationship between probability (p%) of the observed rainfall events that are equal to or greater than a given event and its recurrence interval in year (T) is expressed as

$$T = 100/p \qquad \text{... (4.14)}$$

Therefore, when rainfall data are analysed on certain probability basis, the recurrence interval i.e. frequency is inverse of the probability level.

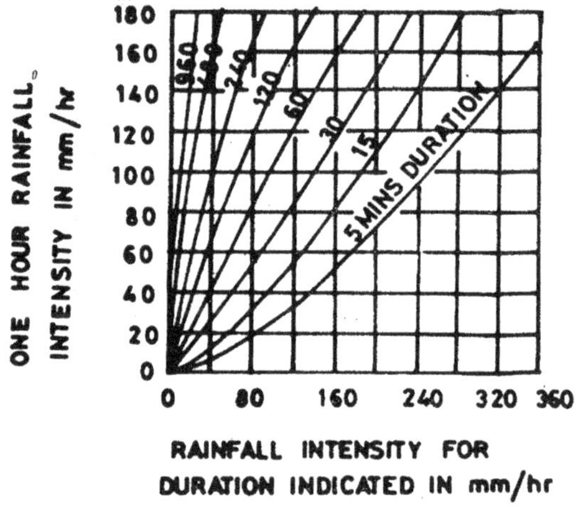

Fig. 4.2. Relation to Convert 1 hour Rainfall Intensity for other Duration

Problem: The maximum rainfall intensity data of a certain location in North India (Hisar) for 20 years are given in Table 4.7. Find out the maximum rainfall intensity that is expected once in 10 years.

Table 4.7: Rainfall intensity data of Hisar (India)

Year	*Rainfall intensity* cmh^{-1}	*Year*	*Rainfall intensity* cmh^{-1}
1971	7.5	1981	4.5
1972	8.5	1982	2.5
1973	5.0	1983	6.0
1974	6.5	1984	7.5
1975	4.5	1985	4.0
1976	3.0	1986	5.5
1977	2.0	1987	2.3
1978	6.5	1988	4.1
1979	3.5	1989	3.8
1980	8.0	1990	1.0

Solution: For estimating the maximum rainfall intensity, the above data are first arranged in the descending order as under:

8.5, 8.0, 7.5, 7.5, 6.5, 6.5, 6.0, 5.5, 5.0, 4.5, 4.5, 4.1, 4.0, 3.8, 3.5, 3.0, 2.5, 2.3, 2.0, 1.0

The using the relation

$$N = T \cdot n,$$

we get $n = N/T = \frac{20}{10} = 2$

The second highest data in the table is the designed maximum (peak) rainfall storm, which is 8.0 cmh^{-1}.

The intensity when estimated by using the relation

$$I = KT^a [t + b]^n$$

and the table 4.7, we get for Hisar

$$I = 5.9160 (10)^{0.1623} (1 + 0.5)^{1.0127}$$
$$= 5.9160 (1.4531) (1.5077)$$
$$= 12.96 \text{ cmh}^{-1}$$

For Jerusalem, the formula for I has been worked out as $I = 15\ T^{0.2}/t^{0.51}$. The intensities for 0.5 and 1.0 hour for 5 years return period were estimated to be 29.5 and 21.6 mm/hour. For Nagev, the average peak flow was expressed as Q = 0.5 (A) 0.67 where Q = discharge (m^3/s) at 20 per cent probability and A = area km^2 and at 50 per cent probability Q = 0.24 (A) 0.67.

The peak intensity when estimated using Fig. 8, given by Tejwani (1975), we get

$$I = 8.0 \text{ cm h}^{-1}.$$

This shows that the estimate of Tejwani is closed to the observed value.

The question which immediately arises is for what probability of recurrence should a water way be designed. This question infact arises for every type of hydraulic structure that carriages flood flows. The problem is one of economics. The cost of designing for a maximum flood in frequent occurrence must be balanced against the damage which might be caused by flood exceeding the capacity of the structure on the average of once in a number of years. This damage may be to the structure itself and to the adjoining property. In case of grassed water ways, the damage due to flood will be very little.

Usually daily rainfall data on 24 hour basis are available at most of the locations or at nearby location and the planning of watershed hydrology is required to be based on these data only, which is of course a crude method, as no other information on the hydrology is available. Such an exercise for a small agricultural watershed (< 500 ha) is illustrated below for Pali watershed (Mahendergarh, Haryana, India).

Design Details

(i) Catchment area and land characteristics

(a) Sandy hills - 20.8 ha
(b) Rockey hills - 18.7 ha
Total - 39.5 ha

(ii) Expected water yield (run off) and design storage capacity

(a) Annual average rainfall - 650 mm
(b) Average monsoon rainfall - 540 mm (July to Sept.)
(c) Average rainfall of the month receiving maximum rainfall - 230 mm (Aug.)
(d) Rainfall of the month receiving maximum rainfall, in 25 years return period (rainfall records for past 50 years are available) - 400 mm

(iii) The total expected runoff from the watershed during monsoon season

$$= \frac{(A_1 C_1 + A_2 C_2) 540}{1000}$$

$$= \frac{(20.8 \times 0.15 + 18.7 \times 0.5) 540}{1000}$$

$$= 7.73 \text{ ha. m}$$

(The values of C_1 and C_2 are taken from Table 4.1)

It is assumed that the storage capacity of the pond shall be repeatedly available for retaining flood water (runoff) on monthly basis due to continuous depletion of water by way of irrigation and domestic use and underground

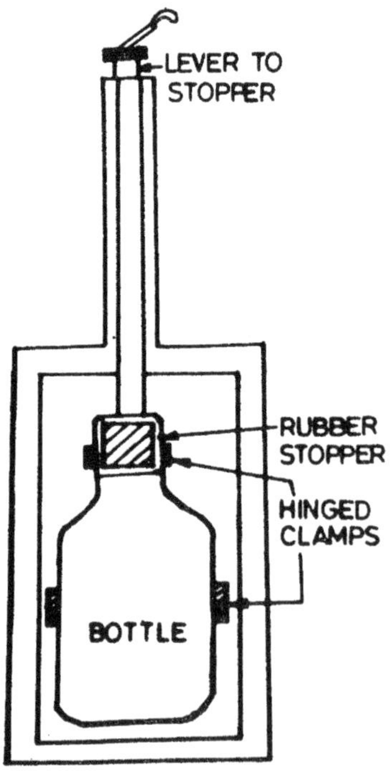

Fig. 4.3. A Suspended Load Bottle Sampler

seepage and evaporation, the dam need not be designed for the entire runoff but it is designed for the month of August, receiving highest monthly rainfall of 400 mm on 25 years return basis. Then the expected runoff yield for 400 mm rainfall is

$$= \frac{(20.8 \times 0.15 + 18.7 \times 0.5)\ 400}{1000}$$

$$= 4.99 \text{ ham}$$

The location of the dam called for having a spillway at a location having elevation 275 m above sea level for which the pond storage worked out to be 5.40 ham.

(iv) Expected peak discharge and the design of spillway
Catchment area of the watershed (A) = 39.5 ha
Weighted runoff coefficient (C) = 0.316
Rainfall intensity (I) for 25 years return period for duration of 1 hour = 8.0 cm h^{-1} (from the Fig. 4.2).

We know that Q= CIA /36

$$= \frac{0.316 \times 8.0 \times 39.5}{36} = 2.77 \text{ m}^3\text{s}^{-1}$$

A trapezoidal spillway with a bottom width 4 m, depth 0.6 m and side slope of 1:1 can take care of the discharge equal to 2.84 $m^3\ s^{-1}$ (using the equation 7.24). Also it will provide good stability. The saddle portion shall be provided with dry stone pitching for protection and attaining proper level. In this design no reduction has been made in the discharge on account of dead storage which is not much and shall not make any material difference in the cost. However a free board of 0.9 m should be provided to the designed height of the dam. Therefore the top level of the dam is

$$275 + 0.6 + 0.9 = 276.5 \text{ m}$$

Details of dam design:

(i)	Bed level of site	– 270.9 m
(ii)	Inlet level of pipe outlet	– 273.0 m
(iii)	Full pond level	– 275.0 m
(iv)	High flood level	– 275.6 m
(v)	Total level of dam	– 276.5 m
(vi)	Total height of dam	– 276.5 – 270.9 = 5.6 m
(vii)	Maximum pondage depth	– 275.0 – 270.9 = 4.1 m
(viii)	Side slope of dam wall	– 3 : 1
(ix)	Top width	– 4.0 m (assumed)
(x)	Storage capacity	
	Live capacity	– 4.4 ham
	Dead storage	– 1.0 ham
		5.4 ham

Pipe outlet: A 0.3 m diameter RCC pipe should be provided to convey the water to down stream.

Estimation of Sedimentation Load

Runoff water carries a lot of sedimentation load. Estimation of this load is important to understand the extent of the soil erosion and to estimate the soil load carried to the reservoir by the stream. Sediments are transported by the process of suspension, saltation and surface creep. The surface creep is the material rolled along the bed while the saltation is the material bouncing along the bed. The sediment transported in suspension is called suspension load. In case of sediment transported by the stream, the saltation and surface creep are combined and is known as the bed load. To get the total sediment load, both the bed load and suspended load are separately determined and added.

For determining the bed load, pressure difference type sampler is used. It consists of a iron sheet closed rectangular channel with a diverging section at one end, a converging section at the another end and a trap in between. The diverging section is placed towards the down stream with converging section at the upstream. The sampler is placed on the stream bed to measure the amount of bed load collected in a given time. It is usually held in position by a rod for shallow streams with a provision of a winch to lower or raise the sampler in the stream. The bed load collected in the sampler is dried and weighed. The dry weight when divided by the time taken for measurement and the width of the sampler gives the rate of bed load movement per unit time per unit width of the stream bed at the point of measurement. With the rate of bed load zero at the banks, a curve showing movement of the load across is constructed from which the total rate of bed load movement for the entire cross section of the stream is calculated.

Sediment measurements are usually made for the suspended load as the bed load measurement is not only difficult, but is also highly variable with time for a given discharge. The suspended load can be determined by a bottle type sample (Fig. 4.3). It consists of a metallic bottle held at one end of a pipe. A lever is attached at the upper end of the pipe and a rubber cork at the lower end. By pressing the lever, the rubber cork is lifted up and the bottle mouth is opened to get the bottle filled with the suspension. By releasing the lever, the bottle is closed and weighed to know the suspended load. For correct observations, the bottle is kept vertical and its mouth is opened at the desired depth (about 60% of the total depth of the stream) of observation for the minimum time required to fill the bottle. Normally five litres of sample is collected from each point of observation.

Once the bed load and suspended load are known, the amount of material transported in a day is given by

$$S = P \times Q \times 86{,}400/1000 \text{ tons} \quad \ldots (4.15)$$

where S is amount of material transported, (t/d), P is amount of material in suspension (kg/m^3), of water and Q is rate of stream flow (m^3/s).

Problem 1: In a watershed there are six rain-gauge stations. During a given rainstorm the rainfall recorded at each of these stations are 7.4, 6.3, 5.3, 8.0, 6.9 and 5.7 cm. Calculate the mean rainfall of the watershed for that rainstorm.

Solution: The arithmetic mean (Rm) of the rainfall data is

$$R_m = \frac{7.4 + 6.3 + 5.3 + 8.0 + 6.9 + 5.7}{6}$$

$$= \frac{39.6}{6} = 6.6 \text{ cm}$$

Problem 2: Estimate the peak rate of runoff for a 25 year return period for a small watershed of 25 ha located at Hisar receiving annual rainfall of 400 mm against PET of 1475 mm. The watershed is having 15 ha under cultivation, 5 ha under forest and 5 ha under pasture. The general slope of the watershed is less than 5 per cent and its soil texture is sandy loam. There is a fall of 5 m in a distance of 700 m which is also equal to the distance from the remotest point to outlet of the watershed.

Solution : The weighted value of runoff coefficient for the entire watershed is

$$R = \frac{15 \times 0.3 + 5 \times 0.1 + 5 \times 0.1}{25}$$

$$= 0.22$$

For L = 700 m and H = 5 m, the time of concentration (T_c, in minutes) is

$$T_c = 0.0195\, K^{0.77}$$

$$= 0.0195 \left[\frac{700^3}{5} \right]^{0.385} \qquad \therefore\; K = \frac{L^{3/2}}{H^{1/2}}$$

$$= 20.27 \text{ minutes}$$

$$= 20.\text{minutes (say)}$$

The rainfall intensity for Hisar (North India) for 25 years return period is 85 mm h^{-1} (from figure 4.1). The rainfall intensity for 20 minutes is calculated by using the equation

$$i_{20} = I\,\frac{T+1}{t+1} = 85\,\frac{[I+1]}{1 + 20/60}$$

$$= 12.75 \text{ cm h}^{-1}$$

$$Q_p = \frac{CIA}{36} = \frac{0.22 \times 12.75 \times 25}{36}$$

$$= 1.95 \text{ cumec}$$

Problem 3: Find out the maximum runoff for a 25 years recurrence (return period) for antecedent soil moisture condition III for a watershed having 35 ha of row crop cultivated good terraced land and 15 ha of good pasture land. The watershed is located at 30^0N latitude and 76^0E longitude. The soil is sandy loam with infiltration rate 5.2 mm h^{-1} and belongs to hydrological soil group B. The time of concentration of the watershed is 0.6 hour.

Solution : Using the Table 4.2, the curve number for cultivated land is 71 while for pasture land it is 61. Therefore, the weighted curve number (C_w) for the entire watershed is

$$C_w = \frac{71 \times 35 + 61 \times 15}{50}$$

$$= 68$$

This curve number corresponds to 85 for antecedent moisture condition III (Table 4.4). The rainfall intensity for 25 year recurrence interval from the figure 4.1 for the location comes out to be 80 mm h^{-1} which for the time of concentration 0.6 hour is

$$I_{0.6} = 80\left[\frac{2}{1 + 0.6}\right] = \frac{160}{1.6} = 100 \text{ mm h}^{-1} = 10.0 \text{ cmh}^{-1}$$

The storage capacity of the watershed(s) is given by equation as

$$S = \frac{1000}{CN} - 10 = \frac{1000}{85} - 10$$

$$= 1.76 \text{ cm}$$

The Runoff from the watersheds is given by the expression

$$Q = \frac{(R - 0.2s)^2}{R + 0.8s} = \frac{(10.0 - 1.76 \times 0.2)}{10.0 + 0.8 \times 1.76}$$

$$= \frac{(9.65)^2}{11.41} = 8.16 \text{ cm}$$

Total quantity of runoff in m^3

$$= (50 \times 100 \times 100)\ \frac{(8.16)}{100}$$

$$= 40.8 \times 10^3 \text{ m}^3$$

Problem 4: The discharge and suspended load concentration recorded in a small stream of 5000 ha watershed is given below. Calculate the volume of runoff, depth of runoff and the sediment load.

Time (h)	0	1	2	3	4	5	6	7	8	9
Discharge (Q) (m^3/s)	0.0	4.5	51.5	38.0	7.0	1.3	0.2	0.07	0.02	0.0
Concentration (C) (kg/m^3)	0.0	1.0	10.0	5.0	1.5	0.4	0.2	0.02	0.02	0.0

Solution:

Runoff = 0.0 + 4.0 + 51.5 + 38.0 + 7.0 + 1.3 + 0.2 + 0.07 + 0.02 + 0.0

= 102.09 m^3/s

= 102.09 × 3600

= 3,67,524 m^3.

Depth of runoff = 367524 × 1000/5000 × 10000

= 7.35 mm.

Sediment load = Q × C

= 4.5 + 515.0 + 190.0 + 10.5 + 0.5 + 0.04

= 720.04 kg/s

= 720.04 × 3600/1000

= 2529 tons.

Problem 5: The hourly discharge and the silt load for a catchment of 100 ha area for one day are listed below. The density of silt is 1.4 Mg/m^3. Calculate the silt load and total production for the day from the catchment.

No.	Time (hours)	Discharge (Q) m^3/s	Silt (S) g/L
1	24	33.5	1.3
2	1	33.5	1.2
3	2	32.0	1.2
4	3	30.0	1.1
5	4	29.0	1.0
6	5	26.5	0.9
7	6	24.5	0.8
8	7	23.0	0.7
9	8	22.0	0.6
10	9	21.0	0.5
11	10	19.0	0.4
12	11	18.5	0.4
13	12	16.0	0.4
14	13	10.0	0.4
15	14	9.0	0.3
16	15	8.8	0.3
17	16	8.6	0.3
18	17	8.0	0.3
19	18	7.6	0.3
20	19	7.2	0.3
21	20	6.9	0.3
22	21	6.5	0.3
23	22	6.1	0.3
24	23	5.0	0.2
25	24	4.8	0.2

Solution: Total silt load (ton) =

Average discharge (m^3/s x Averge silt load (g/L) × 3.6

= [(33.50 + 33.50)/2. (1.30 + 1.20)/2 + ...

.... + (5.0 + 4.8)/2. (0.2 + 0.2)/2] × 3.6

= [(33.5 x 1.25) + ... + (4.9 x 0.2)] x 3.6

= 294.375 x 3.6 = 1059.75 tons.

Volume of silt load/day from the watershed = 1059.75/1.4 x 10000

= 0.0757 ha m/day.

For low and high soil moisture levels, Table 4.4 should be consulted for determining the corresponding curve number.

Chapter — 5

Water Storage

Storage for storing rainfall runoff water to provide critical irrigation is important for dryland agriculture. Any excess rainfall that required for *in-situ* storage in the soil profile needs to be stored properly. In fact need for storage of water arises when the supply of water in a given period is in excess than the water needs, while in the subsequent period it is less than the water demand. Such situations are common in rainfed agriculture. There it is essential to provide an efficient farm pond for storage of rainfall runoff water to make it available at the peak time of demand. An efficient storage retains the water without excessive losses until it has served the intended purpose. Such storage ponds are highly relevant for dryland conditions for improving and stabilizing dryland agriculture and to build up the confidence of the farmers in dryfarming. In fact such farm ponds are not new in dryland areas. The practice of constructing farm ponds for irrigation purposes in south-India and for harvesting roof top rain-water for domestic/drinking purposes in western-India, Australia, Bangladesh etc. is in vogue since ancient times. These ponds serve as potential source for supplemental/critical irrigation, domestic consumption and drinking purpose for human beings as well as for livestock besides serving as ideal loci for acquafish culture and multiplication of biological fertilizers like *azolla*. In fact drought proofing is possible only with the water harvested in farm ponds.

Types of Water Storage

The storage of water are of two types (i) imponding and (ii) off stream. The imponding reservoir is a dam built across the valley to receive the whole of the flow from the catchment area and overflow arrangement for the excess water, must be of sufficient size and strength to accept the pass storm flows past the dam when the reservoir is full. Off stream reservoirs are constructed away from the main stream and the stream is diverted towards the reservoir having a bye-pass to divert the excess water back to the stream when the reservoir is full. Off stream reservoirs are made wholly by excavation or a combination of excavation and embankment. The water storage capacity to earth ratio i.e. the volume of water storage made available by the movement of a given volume of earth is an important consideration for dealing its economical feasibility. The farm ponds are usually feasible on flat terrains to harvest natural or induced runoff.

Reservoir Capacity

Storage capacity of a gully dam/reservoir/farm pond is designed on the basis of water requirement for irrigation purpose, water losses through seepage and evaporation and rainfall-runoff (catchment yield) relationship. The runoff is that portion of the rainfall which is neither absorbed in the ground, stored in the land surface nor evaporated but which flows over land surface. It is influenced by climate and physiographic factors. The climatic factors include rainfall amount, intensity and direction of rain storm, interception by vegetation, evaporation from soil and transpiration by plants. The physiographic factors are characteristics of the watershed and its water channels. The watershed characteristics include size, shape, slope orientation, land use, infiltration characteristics of the soil and topographic conditions. Water channel characteristics are size, shape, cross-section, slope and roughness of the channel bed. A fan-shaped catchment produces more peak rate of run-off than a narrow and long catchment for the same area of the catchment. The amount and peak-rate of run-off can be altered by altering any of these parameters. The storage requirement should be balanced against the amount of run-off expected at a certain probability (confidence) level and irrigation, domestic, animal etc. requirements of water.

Earth Embankments and Dams

From the design and construction points of view, both embankment and dam are synonymous. The two requirements for their design are stability and water tightening. The stability depends upon the type of soil used in the construction, soil underlying the embankment/dam, the slopes of the upstream and downstream faces, the overall width and the measures taken to prevent surface and internal erosion. The resistance of the foundation soil and of the embankment itself to the flow of water is important both from the point of view of seepage loss and stability. The internal seepage results in soil erosion is called piping.

Stability: A foundation for a dam must be strong enough to support the weight of the dam without undue settlement/sliding. Top soil or other soils containing good organic matter is not suitable for making dams/reservoirs. The walls should be slopy towards the reservoir side and level and should be resistant to erosion.

Seepage Control: Seepage through permeable foundation soils which over lay an impervious layer can be controlled by constructing a cut-off wall along the line of the dam. A trench of sufficient width is excavated down to the impermeable layer and is filled with soil rich in clay. The ideal soil is one which contains 20-30 per cent clay and the remainder being well graded sand and gravel. Bentonite (clay) is mixed with permeable soil devoid of clay to make it suitable for dam making. The base and sides of embankment should be treated suitably to reduce seepage. Soils with high clay (>30%) create problem due to shrinkage and swelling of the clay.

Selection of Site for Dam/Embankment

Undulating terrain provides suitable sites for dam construction. At the dam site, sides of gully should be close together to reduce the length of the dam. Stream gradient should be flatter to give more capacity and sides of the storage basin should be widen out upstream. Soils in the reservoir area should have sufficiently impermeable material to prevent seepage losses. In the absence of relatively impervious material (clay soil) over a portion of the area, the previous soil should be treated with sealing materials like clay soil, bentonite etc. The foundation of an earth-fill dam should provide support for the embankment under all conditions of saturation and loading and should have sufficient resistant to seepage. Tentative sites should be tested with 3-4 metre deep bores dug at each corner and in the middle of the site to explore seepage and stability of the foundation to support the embankment. The bores (about 7.5 cm diameter) should be dug in the depth interval of 25 cm with the help of a post-hole-cylindrical auger (Fig. 5.1) and about 0.5 kg of the soil of each depth interval is taken in a cloth bag after mixing the soil thoroughly. The portion

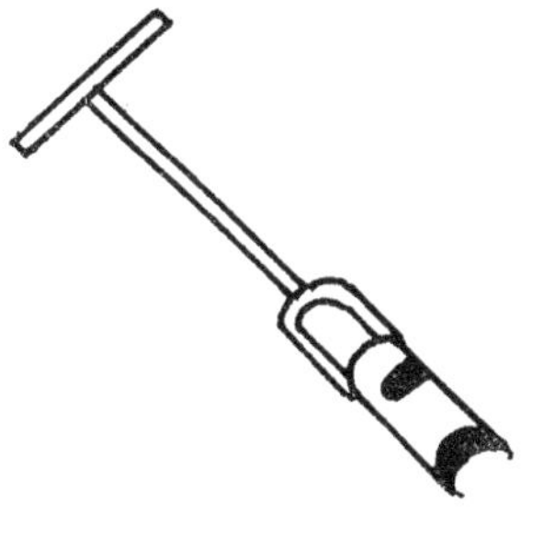

SOIL AUGER

Fig. 5.1. A post hole cylindrical auger

of the various component materials that form a particular soil should be analysed in Soil Testing Laboratory. However estimation about the mechanical composition of the soil can be had as under.

A narrow parallel sides glass bottle of capacity about 1 litre is half filled with the soil material after removing stones and then the remaining bottle is filled with water. The bottle is then shaken well to mix the soil and water. After adding 2 to 3 drops of sodium silicate or about 50 gm of common salts in the bottle, the bottle is stirred well for about 0.5 hour to allow dispersion of the soil in the bottle and the bottle is kept upright for 24 hours to allow the various mechanical fractions of the soil to settle down. In this process, coarse sand will settle down immediately, fine sand within a few seconds and the silt and clay within a few minutes to a few hours. There will be stratification of the soil into closely visible layers from which the approximate proportions can be estimated by measuring the depth of each layer with a measuring scale. If for example, the total depth of the soil in the bottle is 50 cm and that of bottom (coarse sand and gravel), middle (fine sand) and top (slit + clay) layers are 25, 15 and 10 cm respectively, then the proportion of these sized groups of particles are 50, 30 and 20 per cent, respectively. The amount of clay can be estimated by close examination of the top fraction.

Site Prepation: Trees, shrubs, roots and all vegetations should be removed from the entire reservoir site. All top soil (0-15 cm) should be stripped from the embankment site and put to one side for re-use in soiling the slopes and the top of the embankment. The site should be levelled and the exposed subsoil ploughed/disced in a direction along the line (length) of the embankment. This will improve the seal between the embankment and the foundation by creating parallel corrugation at right angles to the line of possible seepage flow.

Constructions

Homogenous Construction: At a site where well graded soil containing 20-30 per cent clay and remainder sand and gravel is available, homogenous construction of the dam is the cheapest. These dams (FIg. 5.2) are constructed

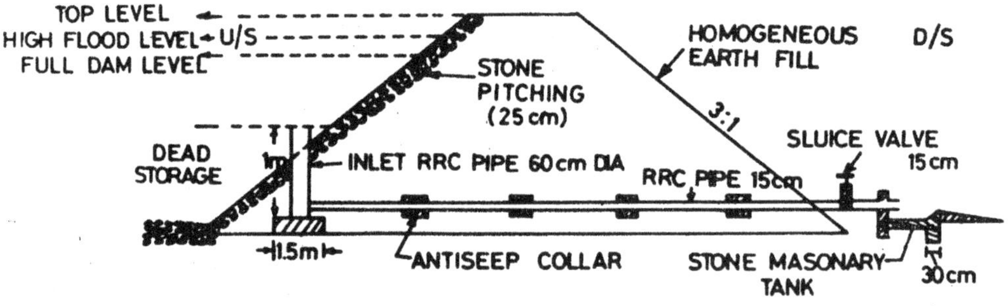

Fig. 5.2. Section of an earthen dam

with uniform and homogenous materials i.e. soil and sand and can be of heights up to 9.0 metres. The dimensions of the dams with and without the impervious layer at just below the keyed foundation is shown in Figure 5.3. The keyed foundation is laid after removing the top soil. In the homogenous construction, the seepage water passing through the embankment emerges on the dry slope and creates the problem of erosion at the toe. To reduce this problem, the soil of the toe is replaced by graded stone filling and by providing horizontal filter drains so that the water passes through the toe without eroding the soil. This measure controls seepage and enables to construct much steeper slopes of the dam and also keeps the phreatic (seepage) line of seepage water within the body of the dam (Fig. 5.4).

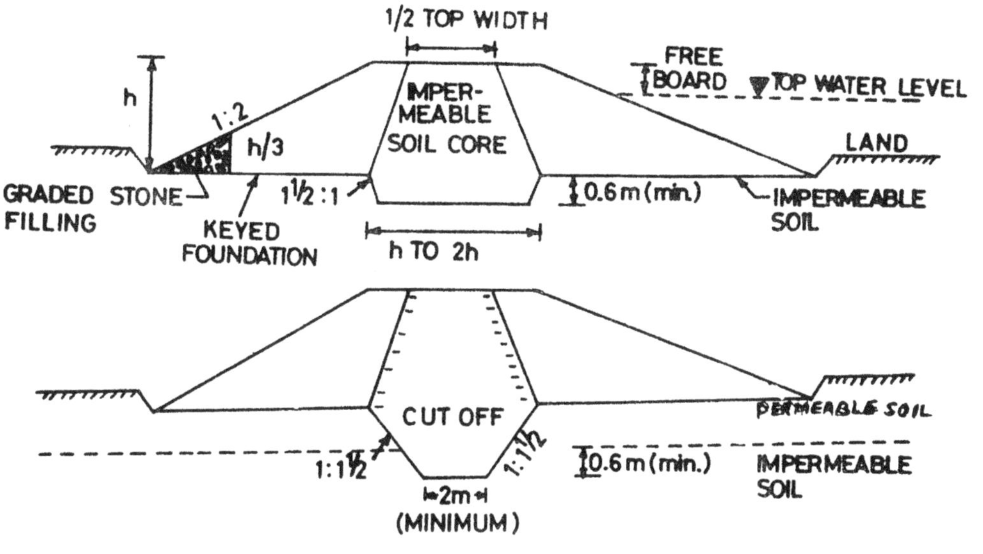

Fig. 5.3. Embankment on impermeable/permeable soils

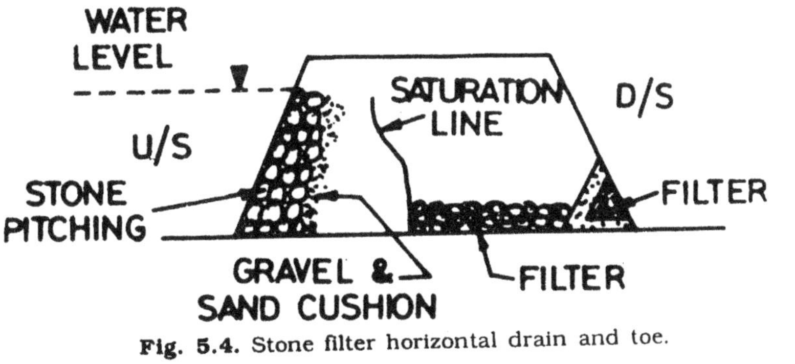

Fig. 5.4. Stone filter horizontal drain and toe.

Zoned and diaphragm construction: Zoned construction should always be useful to provide more stability to the dam where impermeable soil to form seepage barrier i.e. a core (Fig. 5.3) within the embankment, is readily available at the construction site. The permeable soil is then used on the outside to form and stabilize the slopes. The thickness of the core increases with the decrease in clay and the width of the core at the base of the embankment should be twice the height of embankment for soils containing 20-30 per cent clay. The diaphragm (Fig. 5.5) should penetrate atleast 0.6 m below the keyed foundation in dense clay and up to the height of the dam in soils containing 20-30 per cent clay.

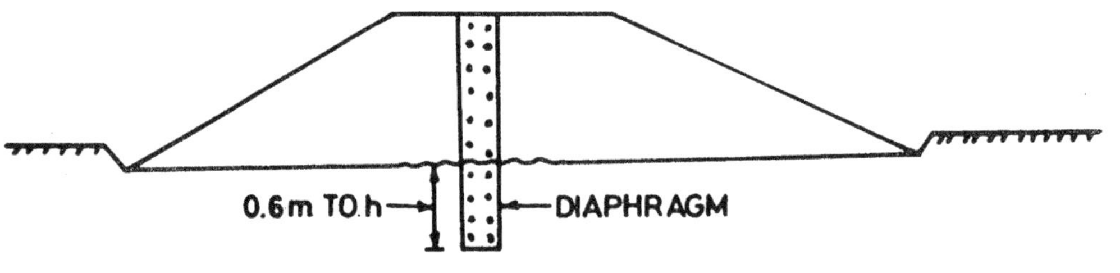

Fig. 5.5. Diaphragm construction of embankment

Blanket Construction : At places suitable impervious soil is not available, the embankment (Fig. 5.6) can be constructed of permeable soil with a water proofing blanket on the wet slope. The blanket may be dense clay, polyethylene or butyl rubber sheeting. The clay blanket must be covered with at least 0.6 m of other material to minimise the liability of cracking and the sheeting blanket must be protected with soil against rain, sun and wind.

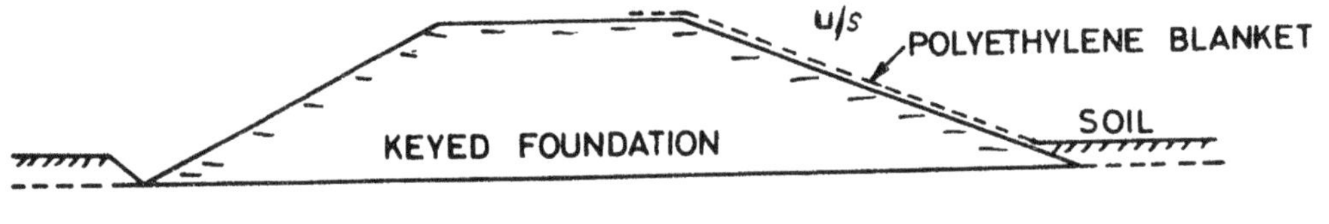

Fig. 5.6. Blanket construction of embankment

Bottom Outlet Pipe : An outlet pipe from the bottom of the reservoir is needed to maintain a residual down stream flow to meet water needs, if any, and to empty the reservoir for cleaning and repairs. This pipe of diameter 0.15 m is laid underneath the embankment/dam in a trench excavated in the undisturbed foundation soil to a depth equal to twice that of pipe diameter, to enable it to withstand any settlement of the embankment and the pressure produce by the full depth of water in the reservoir. Any pipe passing through an embankment must have anti-seepage collars to prevent seepage of water along the wall of the pipe. These collars are made of concrete at least 15 cm thick and have an effective diameter equal to 5 pipe diameters and be spaced 10 pipe diameters apart measured along the line of the pipe. The entrance of the bottom outlet should be located well clear of the upstream toe of the dam and should be 15 cm above the bottom of the reservoir to allow for silting. A trash screen over the end of the pipe is necessary to prevent submerged debris entering the pipe and the floor of the inlet chamber should be at least 15 cm below the invert of the pipe. The pipe should be of sufficient strength of cast iron connected with flexible joints and be fitted with a control valve at the upstream end to control the flow of water through it. This would mean to have an access bridge to a structure above the valve.

Method of Constructing Embankment

The embankment should be constructed by placing the selected soil in successive layers, each layer (0.15 m thick) being continuous over the full width and length of the embankment. As each layer is placed it should be fully consolidated by means of plant at the optimum soil moisture (about 2/3 of the field capacity moisture) content of the soil. The plant may be a heavily loaded rubber wheeled vehicle or a heavy roller. For smaller reservoirs, tractor and scraper earth moving equipment is commonly used for embankment construction. The soil is scraped from the reservoir floor and deposited in layers on the embankment site which gets compacted by its heavy wheels if maintained at the optimum moisture content. As the embankment is raised, its slopes should be roughly trimmed to the correct grade and on completion of the embankment, they should be finally graded. The top of the embankment should be slightly sloped to shed rainfall and to prevent ponding. The construction height of the dam should be about 10 per cent more than the designed height to allow any settlement. The exposed surface of the embankment should be covered by a layer of top soil (0.15 m thick) and seeded with a suitable grass to prevent soil erosion. Stones pitching of the wet slope may be done if the erosion problem is intense. The side slopes of upstream and downstream should be $1 : 2\frac{1}{2}$ and 1:2, respectively.

If the reservoir is excavated below the ground level, the side slopes should not be steeper than $1 : 2\frac{1}{2}$. The minimum core width varies from height of the dam to twice the height for soils having clay contents more than 30 per cent (dense clay) and between 20-30 per cent respectively. The sides of all pits, holes and the cut-off trench, if used should be sloped at about 1.5:1 (horizontal : vertical) to provide a satisfactory grip with the fill.

Estimation of Diamension of Embankment

The capacity of a reservoir is decided based on the quantum of runoff water available, evaporation and seepage losses and water requirement of the area, for irrigation and other domestic purposes. For a given reservoir capacity the required maximum height of dam and its corresponding water to earth ratio can be worked out from Figs. 5.7 and 5.8. For example a reservior of 16,000 m^3 net capacity is to be constructed in a valley which has an average bed slope of 1 in 50 and average side slope of 1 in 20. Then from the Figure 5.7 the intersection of the 1 in 20 valley side slope line and the reservoir capacity 16,000 m^3 is transferred downward to meet the 1 in 50 valley bed slope line.

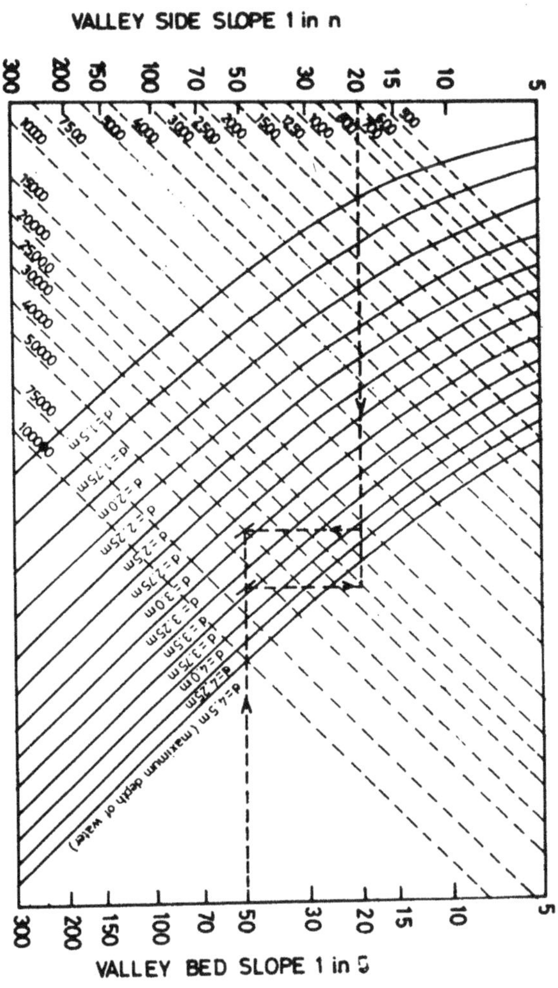

Fig. 5.7. Storage capacity and maximum depth of water

The point of interaction gives a water depth of 3.2 m. Assuming 0.5 metre as losses, the actual depth of water comes to 3.7 metres and the corresponding actual reservoir capacity comes out to be 25,000 cubic metres. The maximum final height of the dam comes out to be 3.7 plus 0.9 (free board) equal to 4.6 metres. The water to earth ratio from the Figure 5.8 corresponding to 3.7 metre height is 6.6 and the amount of earth work is about 4,000 m^3.

Spillways : To ensure that the water does not flow over the embankment, spillways are provided. The spillway must be of sufficient capacity, strength and durability to deal with the maximum storm flow which is most likely to occur. Usually a primary spillway and an emergency spillway are provided. The primary spillway takes care of the most common overflow and reduces the operation frequency of the emergency spillway which disposes off the discharge taking place once in 25 years for small gully dams. The primary spillway is designed to take care of about 25 per cent of the design storm flow of an emergency spillway which is calculated based on the Rational formula (Eq. 4.7) and the rainfall intensity of 25 years return period. The size of the spillway can be computed from the following formula:

$$Q_p = 1.53\ LD^{3/2} \qquad \ldots (5.1)$$

where Q_p = peak discharge in cumec, L = spillway width in metre and D= difference in elevation between spillway crest and the height of the dam in metre.

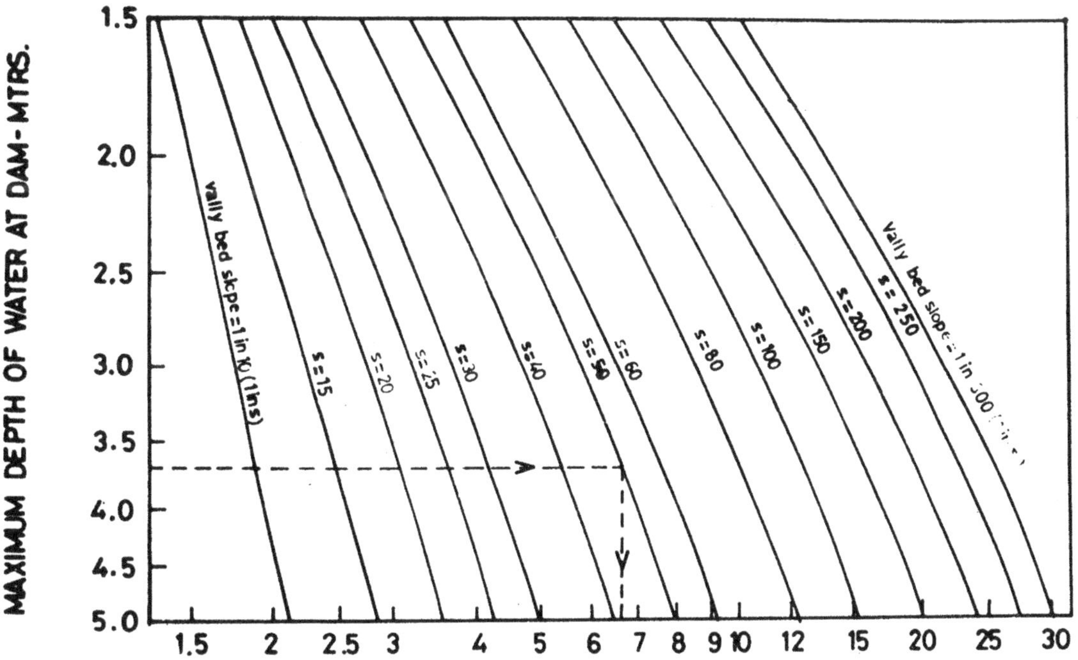

Fig. 5.8. Water to earth ratio of a reservoir

Drop type spillways are generally used on farm reservoir (Fig. 5.9). This type of spillway may be used for draining the pond as well as to supply the water for irrigation. A pipe barrel (diameter 0.15 m) located under the embankment with a riser connected to the upstream end works as spillway. The elevation of the riser is kept at pool water level. Valves and gates are provided in the upstream barrel to regulate the water supply. Hooded inlet spillways are installed to increase the hydraulic efficiency. Sometimes overflow spillways in which water is allowed to drop freely from the crest (weir) are provided. The downstream face of the spillway may be kept vertical or inclined. In some cases the crest of the spillway is extended in the shape of an over hanging lip to keep small discharges away from the straight section of the dam/weir. The water is allowed to fall freely under gravity. For dissipating the energy in some cases, a low height bund is constructed on the downstream side for creating a pool. Principal spillways are constructed with permanent material to provide flood protection. Normally emergency spillways are an earth or vegetative channel designed to discharge flow in excess of the principal spillway. On small farm dams, vegetative spillways are generally used for handling excess runoff. These spillways are designed to discharge peak runoff at a non-erosive velocity to a safe point of release.

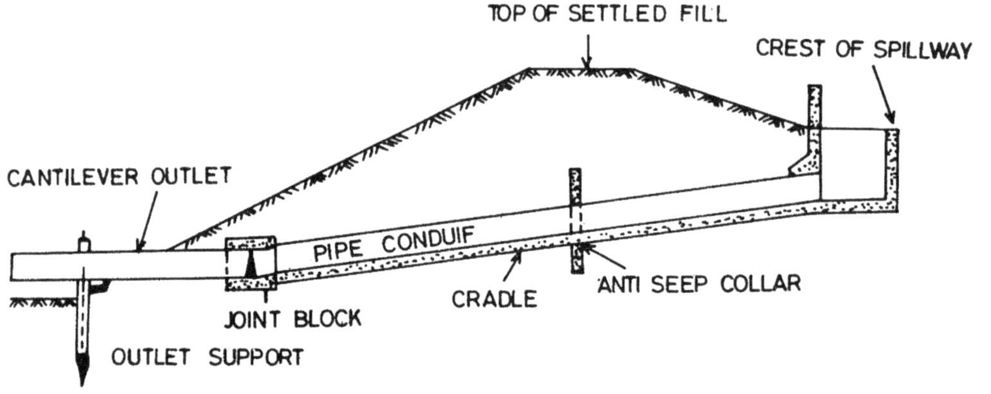

Fig. 5.9. A drop inlet spillway with cantilever

Top Width: The top width of earthen dams should not be less than 2 metres for dam height upto 3 m and it should increase by a metre for each extra 3 m height. The generally accepted formula for crest width is given by

$$W = 1.67 \sqrt{H} \quad \text{... (5.2)}$$

where W = crest width (m), and H = height of crest of embankment above gully bed (m).

Free Board: Free board is the additional height of dam provided to mitigate wave action and runoff from storms greater than the design frequency of the spilling discharge. It is the distance between the elevation of water surface in reservoir when spillway is discharging at the designed depth and the elevation at the top of the dam after settlement. Free board should not be less than 0.5 m for small gully dam.

Freeboard is also defined as the height of crest of the wall above the spillway level and it includes (i) the designed depth of spillway flow-normally 0.5 m (ii) wave height, which can be calculated from Hawksley's formula :

$$H = 0.0456 \sqrt{F} \qquad \text{... (5.3)}$$

where H = wave height (m) and F = fetch length (m), and (iii) an arbitrary allowance for unevenness in crest (≈ 0.1 m). The fetch distance is the longest exposed water surface of the storage reservoir.

Allowance for Settlement: A settlement allowance of more than 5 per cent should be given to the designed height of dam for taking care of consolidation of fill material and foundation material due to weight of dam and increased moisture during storage.

Problem 1. A watershed reservoir is expected to discharge overflow at the rate of 3.0 cumec. If the height of the reservoir is 1.0 m above the crest height of the spillway, find out the width of spillway required for safe disposal of the discharge.

Solution : We know from equation (5.1)

$$Q = 1.53L\,D^{3/2}$$

$$\text{or} \quad 3.0 = L\,(1)^{3/2} \times 1.53$$

$$\text{or} \quad L = \frac{3.0}{1.53} \approx 2 \quad \text{metres}$$

Ans.

Problem 2: Determine the crest width of the dam of heights 4.84 metres.

Solution : We know from equation (5.2)

$$W = 1.67\sqrt{H}$$

$$W = 1.67\sqrt{4.84}$$

$$= 1.67 \times 2.2$$

$$= 3.7674 \quad \text{metres}$$

Ans.

Determination of optimum soil moisture content for maximum bulk density

It is that moisture content of a soil at which the soil could be compacted to the maximum bulk density, when campacted at a given compacting pressure. In order to determine the optimum moisture content, a standard test known as the Pocter density test is conducted in a laboratory. In the test, the soil from the dam site which is to be used for constructing the earthen dam, is placed in a container (mold-15 cm diameter) at different moisture contents (5, 10, 20, 25 and 30% on weight basis) and at each moisture content of the soil, it is compacted with a known pressure (5 kg hammer). Usually the pressure corresponding to the pressure of the equipment to be used (roller 1.5 t) for compacting the soil of the dam wall is selected for laboratory compaction test. The hammer is dropped from a certain height (≈ 0.45 m) for a definite number (50) of times. The moisture content and density of the compacted soil is determined after each compaction test corresponding a definite moisture content and a graph as shown in the Figure 5.10 is drawn. From the graph, the optimum moisture content and the bulk density at maximum compaction is determined. The maximum bulk density should be about 2.0 Mg m^{-3}.

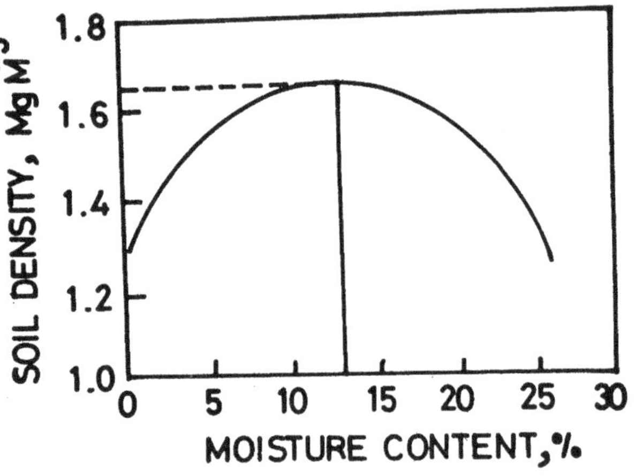

Fig. 5.10. Bulk density moisture content relationship of a soil

Offstream Reservoir/Pond

Offstream reservoirs are constructed to avoid the hazards of dam building and the need of providing large capacity overflows. Gully dam storage or hillside storage sites are not available on flat land and thus offstream reservoirs (also called excavated tanks) are the only alternative for storing harvested water. The excavated tank is generally fed by runoff in rainfed areas and be located in a natural depression and to one side of water way, if possible, so that the runoff water could be diverted and stored properly. Impervious layers of clay or silty clay are very suitable for the site of tank. If possible, the reservoir depth should be deep, limited to the availability and pumping conditions, so as to reduce the containment area and thus minimizing seepage and evaporation losses.

Normally, surface runoff enters into an excavated reservoir through a channel rather than through a board shallow stream and its over-fall from the channel to the bottom of reservoir may create erosion. Also the side slopes of the embankment and channel at inlet are eroded due to scouring and pond capacity is thereby reduced by sedimentation. A suitable structure must be provided at the inlet for safe conveyance of water in the reservoir. Pipe or chute type inlet may be constructed according to inlet conditions. Their size may be decided on the basis of peak discharge of catchment. For off channel reservoir, the size of the inlet structure may depend on the designed runoff which is diverted from the main stream. Sometimes off-stream reservoirs are filled by pumping and the inlet in this case will depend on the pump capacity. A silt trap is essential to check silt into reservoir where runoff water carries considerable amount of silt. A grass filter can be provided in the drainway immediately before inlet to check silt load. The width of the drainway near the inlet should be more than the inlet width and should be grassed to create additional roughness and to further reduce flow velocity.

Site of Reservoir

The following points should be kept in view while selecting the pond site.

(i) ponds should be away from the sites where the chances of contamination are existing e.g., sewage water, polluted water etc.
(ii) at the upstream side about 30 metre surface should be vegetated to work as grassed water way to check silt load into the pond,
(iii) good fill or lining material is available in the vicinity.
(iv) site should have largest storage volume per unit volume of earth work exposed, and
(v) surface area of pond should be minimum.

Size of Ponds: Farm pond size is decided on the basis of total requirement of water for irrigation and the amount of runoff expected from the area and utilization rate of the stored water. The water requirement for crops (Table 8.1), domestic and livestock is reported in Appendix 6. An allowance of 20 per cent is always given to the pond capacity towards losses and uprecedented flood situation.

The dimensions required to construct a reservoir of any given capacity, so that the soil excavated from the reservoir floor area will be just sufficient in amount to form the embankment, can be had from the Figure 5.11. To use this figure, the net reservoir capacity required must first be determined. For example, the depth of water and capacity, after allowing the losses, are required for a reservoir having net capacity of 16,000 m^3 and a perimeter of 370 meters. The 370 m point on the figure 5.11 is extended across to meet the 16,000 m^3 point extended upwards. The point of intersection of these two lines gives the depth of water 2.5 metres. To allow for losses in the case of storage, the depth of water is increased by 0.5 m to 3.0 m and the corresponding actual capacity of the pond is found by extending the 370 m line across the figure to meet the 3.0 metre depth of water line, which gives an actual capacity of 18,000 cubic metres. Figure 5.12 gives the average depth of excavation over the reservoir floor area needed to produce sufficient soil for the embankment of square reservoir of given perimeter and depth of water can also be used with reasonable accuracy for rectangular reservoirs with a length to width ration of upto 1.5. The average depth of balanced excavation required for a reservoir

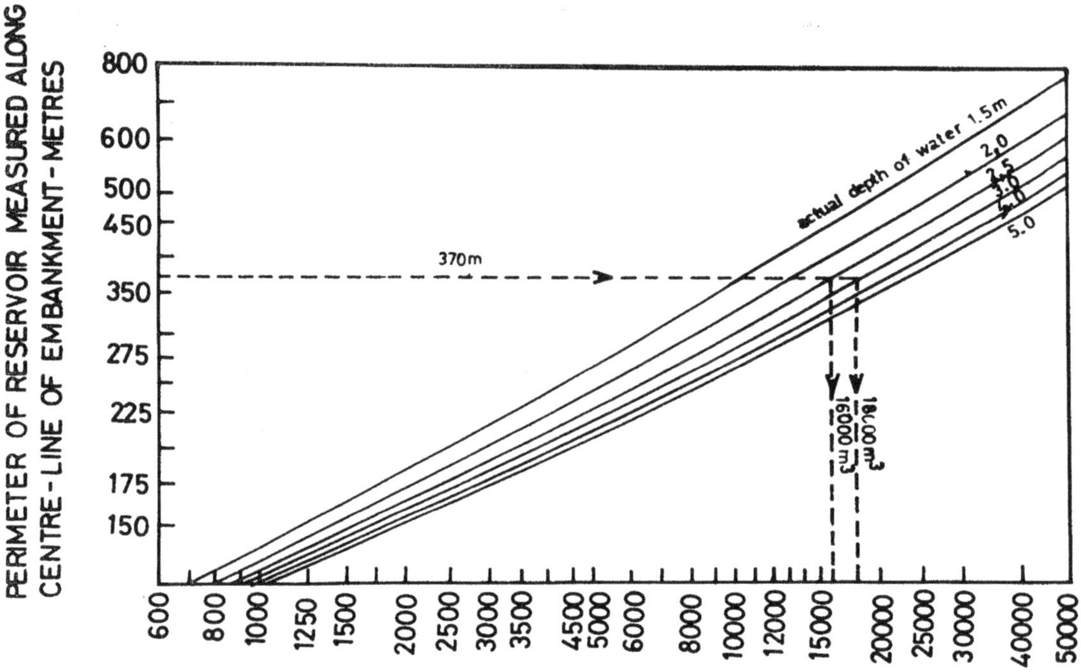

Fig. 5.11. Capacity and depth of water for a reservoir

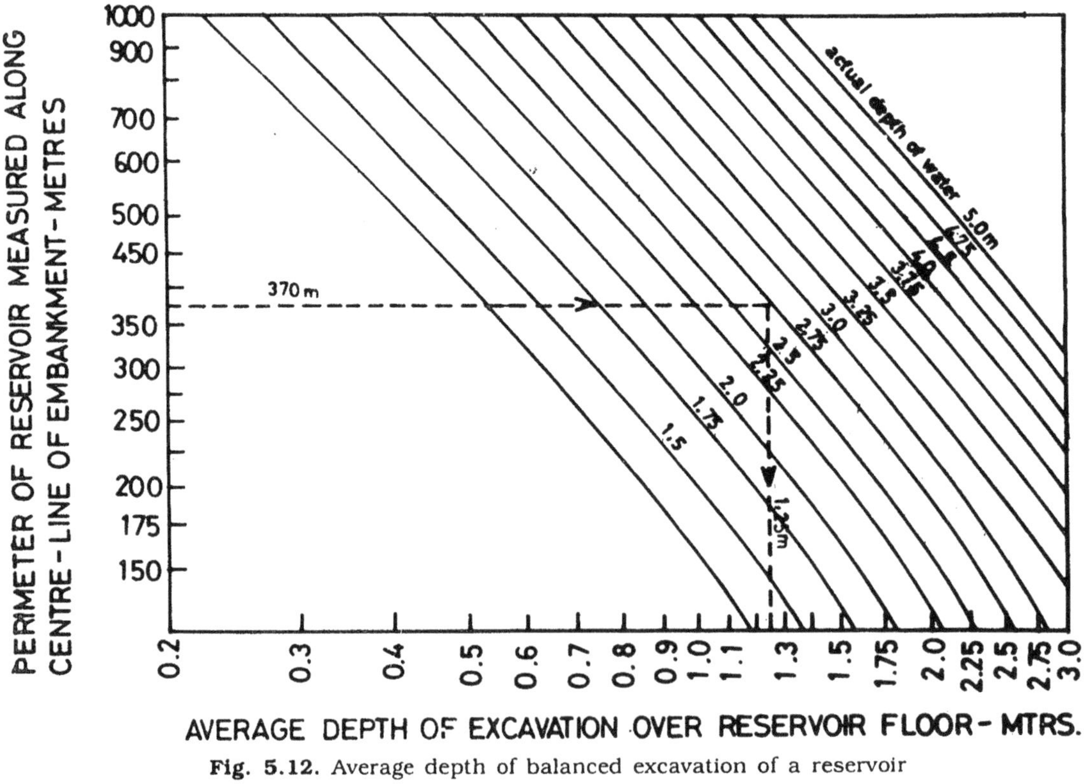

Fig. 5.12. Average depth of balanced excavation of a reservoir

of perimeter 370 m and depth of water line 3.0 m comes out to be 1.25 metres from the Figure 5.12. The water to earth ratio of square reservoirs of given perimeter and depth of water can be estimated from Figure 5.13. The figure is reasonably accurate for rectangular reservoirs with a length to width ratio of upto 1.5. The water to earth ratio for a reservoir of 370 m perimeter and 3.0 m depth of water line comes out to be 2.6.

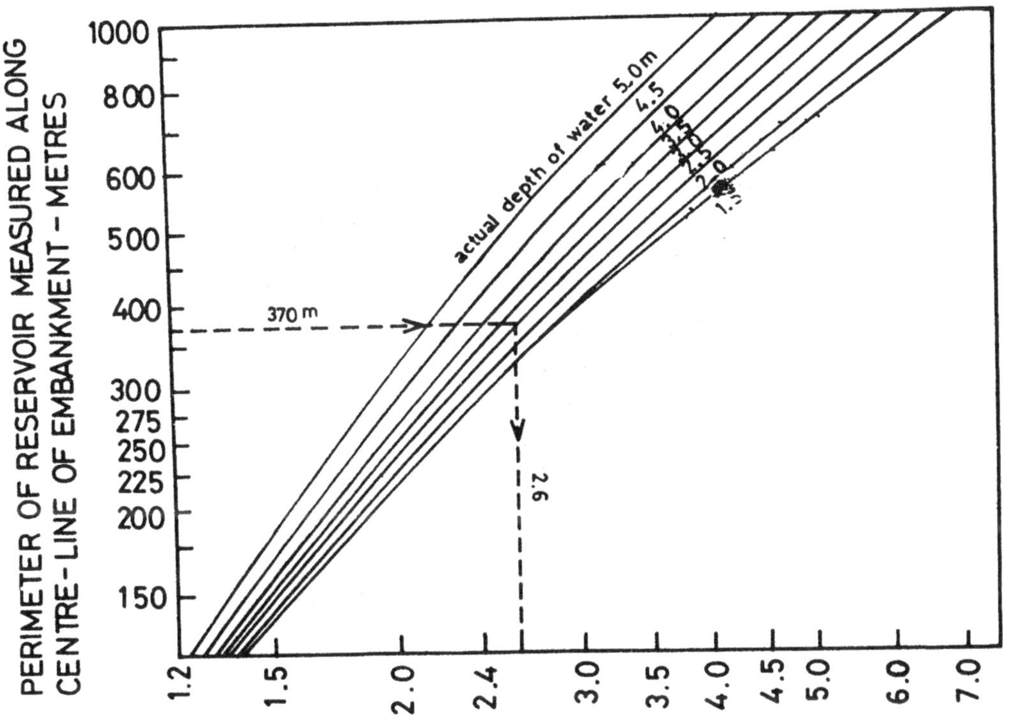

Fig. 5.13. Water to earth ratio of a square reservoir

Seepage Control

Seepage is one of the main losses from storage reservoirs, especially in the regions where soil is too permeable. Proper site investigation at the planning stage is essential to select a site having a relatively impermeable soil. However, the need for water and other factors relating to site selection may be so important to justify the selection of a permeable site. Seepage increases with the wetted area of a reservoir and special consideration during design should be given to minimize the wetted (containment) area per unit of storage capacity. This may be achieved by steepening the sides of reservoirs and increasing the depth in relation to the surface area.

Seepage losses from reservoirs tend to decrease with time due to silt and clay depositions. A newly dug pond may take about 8 years to stabilize the seepage rate at low levels. Sealing of reservoirs is required to effectively control the seepage losses. Some of the seepage control methods are described below:

Particle Size Distribution: Seepage loss from a reservoir can be reduced effectively by mixing large particles (atleast 15 times larger than the weighted particle size of soil) into the soil. By adding large aggregates, the amount of other sealing material may also be reduced. If clay is used to reduce seepage, its amount may be reduced to half by adding larger aggregates.

Puddling: Trampling by animals in newly dug ponds induces puddled conditions and checks seepage losses.

Sealing by Compaction: Compaction of the reservoir bottom/side soils at optimum moisture content can reduce larger soil pores and seepage loss. A thin layer of impervious soil should be spread and mixed with 20 cm depth and the loose soil should be compacted with a heavy duty (1.5 ton) roller at optimum soil moisture condition to achieve good results.

Clay Blanket: Storage areas that have coarse grained permeable soil may be sealed by adding a blanket cover of clay soil. The thickness of the blanket varies with the depth of water to be impounded. A tank of 3 metre water depth should be sealed with at least 25 cm of clay blanket. A gravel cover of 30-40 cm should be placed over the blanket to protect it from cracking. Side slopes of the reservoir walls should be flatter to prevent the blanket from sliding under gravity.

Bentonite: Bentonite (a clay) can absorb several times more water and swells 8 to 15 times its volume. The seepage loss in a well graded soil may be reduced by mixing and compacting bentonite in the surface soil of reservoir. The bentonite should have sufficient exchangeable sodium for its dispersion. The effectiveness of bentonite reduces due to its flocculation due to wetting and drying of reservoir and removal of surface soil-layer by erosion.

Soil Deflocculents: Soil deflocculents like sodium carbonate, sodium chloride, sodium tripolyphosphate etc. can be used for reducing permeability of storage reservoirs. The effectiveness ($\approx$ 18 months) of sodium carbonate is more than the others. The amount of sodium carbonate required for this purpose can be worked out by the following equation:

$$Na_2CO_3 = 0.0428\ DA\ (0.15\ CES - ES) \qquad \ldots (5.4)$$

where Na_2CO_3 = sodium carbonate, g,

D = depth of soil treated; cm,

A = area to be treated; m^2,

CEC = cation exchange capacity, me/100 g, and

ES = exchangeable sodium, me/100 g of soil.

The applied sodium of Na_2CO_3 reduces Ca, Mg from the clay complex and it takes 2-3 years for effective replacement. The sodium clay causes dispersion of the soil and sealing of larger soil pores which reduce seepage. The soil should have clay percentage more than 15 and the thickness of Na_2CO_3 should not be less than 3 cm at each application.

Soil Cement: Soil cement lining is very effective in controlling seepage from reservoir. In a well graded sandy soil, 7-12 per cent cement should, be mixed (on volume basis) in the top 7.5-15 cm soil layer. After wetting the mixture, it should be compacted at optimum moisture to avoid cracking of the soil.

Concrete Lining: Concrete line is costly but is most effective and is used for tanks for drinking water and industrial purpose.

Brick, Tiles or Stone Lining: Reservoirs may be lined with bricks, tiles and stones according to availability of material. The gaps between bricks/tiles/stones may be filled with cement sand plastering. Lining with cement is very effective to control seepage. However, it is also a costly measure of seepage control.

Chapter — 6

Land Levelling and Earth-work Estimation

Agricultural land used for cultivation requires an even surface for crop production. Undulating lands need land levelling for making it suitable for cultivation and efficient use of limited water available for irrigation. Land levelling involves grading and smoothening of land surface for efficient and uniform distribution of water over land surface as well in soil in the soil profile, both under rainfed and irrigated conditions. Land grading consists of reshaping the land surface to a desired grade. It removes humps and depressions interfering with the flow of water on the land surface. Smoothening of levelled land eliminates, minor undulations and facilitates surface drainage of water. Land grading in irrigated agriculture helps in uniform application of water, better regulations, saving of irrigation and leaching of salts. Under rainfed conditions land levelling helps in uniform soil and water conservation. It is not necessary to bring a land to perfect level. Usually a moderate grade is required for surface drainage of water. For border and furrow methods of irrigation there is a need of land levelling while for drip and sprinkler methods there is not much requirement of land levelling. Depending upon the topography, land levelling requires displacement of a lot of soil, sometimes over a large distance and over a large area. This involves a lot of earth work.

The land grading required for an area depends upon the topography of the area, soil type, soil depth, crops to be grown, source of water supply and method of irrigation. The depth of the soil puts a rider on the depth of cut in land levelling operations as sub-soil (unsuitable for cultivation) should not get exposed and soil fertility should not be disturbed drastically. However, there is no such limitations for sand dunes. For an average soil, removal of top soil should not exceed 2.5 cm/annum. Usually the terms land grading and land levelling are interchangeably used to mean the same thing.

Planning for Land Levelling

Prior to undertaking any levelling work of a land, heavy vegetation and plant residues should be removed and major undulations should be flattened as far as possible. The planning for the land levelling should take into consideration the entire farm. Information about the soil textures, soil profile, existing field boundaries, water courses, sources of irrigation etc. should be obtained. A topography map of the area with the levels marked is required for calculating the earth-work involved in grading and cost estimates. A plan of the land area is required which if not available should be prepared by surveying (chain) the area. Such a plan should be drawn on a graph paper to a suitable scale. The whole area should be drawn on a graph paper to a suitable scale. The land should be divided into squares (grids) with the help of ranging rods and cross-staff. The side of each grid may be 30 m × 30 m or 15 m × 15 m or 8 m × 8 m depending upon the size of the field, topography and accuracy in work needed. For convenience in making the grids, square corner of the field i.e. one which is right angle is selected. If such a corner is not available, two base lines are drawn at right angles and their position is marked on the field as well as on the plan. The first line should be parallel to the base line at a distance of half the proposed grid size. Pegs are fixed on this line at the prescribed grid distance. Similarly, pegs are fixed on the line perpendicular to this line and at all grid points. By this procedure each peg will be at the centre of each grid and the level of each peg will represent the level of the entire square. After marking the points, their elevation readings are taken with the help of a Dumpy level and reduced levels are calculated with respect to a bench-mark. The levels taken at the corner of the grids are used for calculating the earth-work involved in the grading. After the contour plan of the entire area is ready the whole area is divided into fields in accordance with topographic situations such that each field can be levelled to the best advantages, with minimum earth-work involvement. The desirable land grade for different types of soils are as under:

Texture	*Grade %*
Heavy soils	< 0.05
Medium soils	0.05-0.20
Light soils	0.20-0.60

These slopes may be modified to suit the designed runoff coefficient values for a given land area.

Land Levelling Calculations

There are several methods for land levelling calculations. However, the plane method, in which the land surface has uniform field slope and uniform cross-section, is simple, precise and is commonly used. The average elevation of the field is determined and is assigned to the centroid of the area. The centroid is located by taking moments about the two axes (X and Y). Any plane passing through the centroid is expected to produce equal volumes of cuts and fills of the land. Such a plane could be described mathematically by an expression:

$$E = a + S_x X + S_y Y \quad \text{... (6.1)}$$

where E = elevation of any point in the field, a = elevation of the datum point (or origin) and S_x and S_y are land slopes along X and Y directions, respectively. The new elevations of all the stations are worked out with respect to a station which is nearest to the centroid and has nearly same elevation as that of the centroid.

Calculations for locating centroid

Distance parallel to X-axis

Line	*Distance, m*	*No. of station*	*Product*
A	30	7	210
B	60	7	420
C	90	7	630
D	120	7	840
E	150	5	750
		33	2850

Distance parallel to Y-axis

Line	*Distance, m*	*No. of station*	*Product*
1	0	5	0
2	30	5	150
3	60	5	300
4	90	5	450
5	120	5	600
6	150	4	600
7	180	4	720
		33	2820

The distance of centroid from Y-axis

$$= \frac{2820}{33} = 85.45 \text{ m}$$

Example: Determine the volumes of cuts and fills and the cut-fill ratio for a field whose topographic elevations are given in Figure 6.1, to have (i) a levelled field and (ii) a graded field of 0.2 per cent slope along both X and Y axes, respectively.

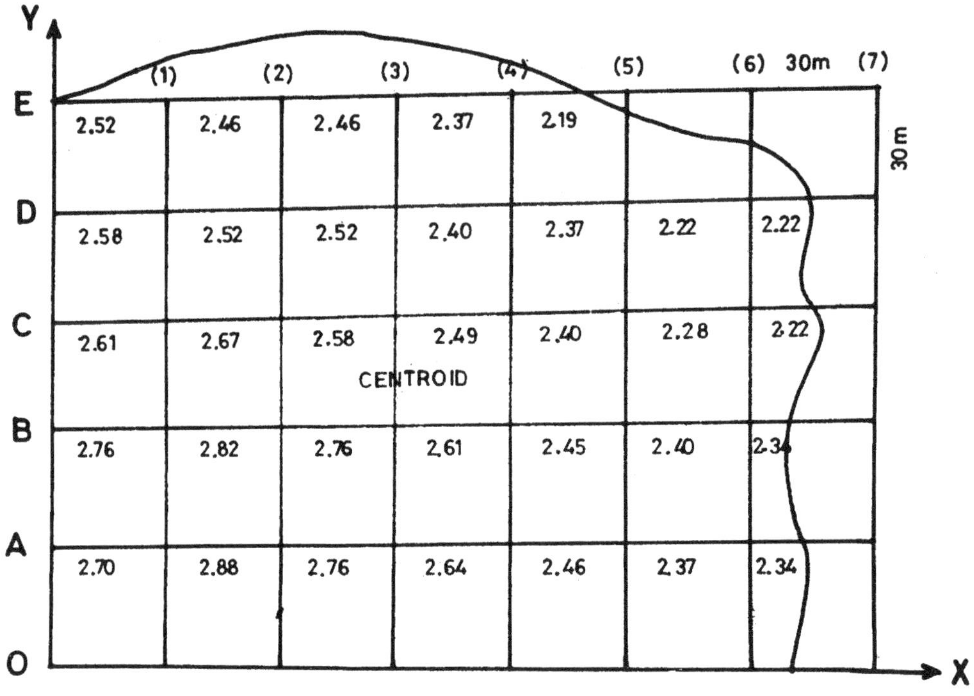

Fig. 6.1. Elevation of a field at different grid points

The minimum level in the whole area is 2.19 m. If the area is cut at this level, the amount of cut at each station will be as follows:

	(1)	(2)	(3)	(4)	(5)	(6)	(7)
E	0.33	0.27	0.27	0.18	0.00	–	–
D	0.39	0.33	0.33	0.21	0.18	0.03	0.03
C	0.42	0.48	0.39	0.30	0.21	0.09	0.03
B	0.57	0.63	0.57	0.42	0.30	0.21	0.15
A	0.51	0.69	0.57	0.45	0.27	0.18	0.15
	2.22	2.40	2.13	1.56	0.96	0.51	0.36=10.14

Since about 90 percent of the cut soil can be used in filling, the cuts should be about 10 percent more than the fills. The cuts in this case is about 10 per cent more than the fills, the levelling at 2.49 m elevation is desirable.

For 0.2 per cent slope the elevation of the station near the centroid, which is station 4 on C-line (elevation 2.49 m) is taken as datum and elevation of other stations for 0.2 percent slope is evaluated by adding or sub-tracting 0.06 m to the elevation of all the stations right or left to the datum station. The new elevations as worked out by the equation (6.1) are as under:

	(1)	(2)	(3)	(4)	(5)	(6)	(7)
E	2.56	2.49	2.43	2.37	–	–	–
D	2.61	2.55	2.49	2.43	2.37	2.31	2.25
C	2.67	2.61	2.55	2.49	2.43	2.37	2.31
B	2.73	2.67	2.61	2.55	2.49	2.43	2.37
A	2.79	2.73	2.67	2.61	2.55	2.49	2.43

Amount of earth work for levelling at 2.19 m

$$= 30 \times 30 \times \sum \text{cut}$$

$$= 30 \times 30 \times 10.14 = 9126 \text{ m}^3$$

If this amount of earth is spread over 33 stations the ground level will rise by

$$= \frac{30 \times 30 \times 10.14}{30 \times 30 \times 33} = 0.3072 \approx 0.31 \text{ m}.$$

The final ground level will be $2.19 + 0.31 = 2.50$ m. If the cuts and fills are calculated with this ground level, they will balance each other as mentioned below.

	(1)		(2)		(3)		(4)		(5)		(6)		(7)	
	C	F	C	F	C	F	C	F	C	F	C	F	C	F
E	0.02	–	–	0.04	–	0.04	–	0.13	–	0.31	–	–	–	–
D	0.08	–	0.02	–	0.02	–	–	0.10	–	0.13	–	0.28	–	0.28
C	0.11	–	0.17	–	0.08	–	–	0.01	–	0.10	–	0.22	–	0.28
B	0.26	–	0.32	–	0.26	–	0.11	–	–	0.01	–	0.10	–	0.16
A	0.20	–	0.38	–	0.26	–	0.14	–	–	0.04	–	0.13	–	0.16
	0.67	–	0.89	0.04	0.62	0.04	0.25	0.24	–	0.59	–	0.73	–	0.88

Total cuts = 2.43

Total fills = 2.52

Since cut is less than fill, it is not desirable as it would involve borrowing the earth from outside the area. Therefore it is proper to assume the new level.

Let us assume the new level at 2.49 m instead of 2.50 m. Then the new cuts and fills at each station (grid point) are as under.

	C	F	C	F	C	F	C	F	C	F	C	F	C	F
E	0.03	–	–	0.03	–	0.03	–	0.12	–	0.30	–	–	–	–
D	0.09	–	0.03	–	0.03	–	–	0.09	–	0.12	–	0.27	–	0.27
C	0.12	–	0.18	–	0.09	–	0.00	–	–	0.09	–	0.21	–	0.27
B	0.27	–	0.33	–	0.27	–	0.12	–	–	0.00	–	0.09	–	0.15
A	0.21	–	0.39	–	0.27	–	0.15	–	–	0.03	–	0.12	–	0.15
	0.72	–	0.93	0.03	0.66	0:03	0.27	0.21	–	0.54	–	0.69	–	0.84

$$\sum \text{Cuts} = 2.58$$

$$\sum \text{Fills} = 2.34$$

Since cuts are more than fills it is desirable

Earthwork Estimation

After the cuts and fills of each grid point are calculated the earthwork can be estimated by the expression:

$$V_c = \frac{A\,(\sum \text{cuts})}{4\,(\sum \text{cuts} + \sum \text{Fills})} \quad \ldots (6.2)$$

where A = area of the grid. For computing the volume of fill V_f, $(\sum \text{cuts})^2$ in the numerator of the above equation is replaced by $(\sum \text{Fills})^2$. If the earth work calculations are for a part of the grid, then A should be equal to the area of that part of grid.

For example in the above mentioned problem

$$V_c = \frac{30 \times 30 \times 2.58^2}{4\,(2.58 + 2.34)} = 304.4 \text{ m}^3$$

$$= \frac{30 \times 30 \times 2.34^2}{4\,(2.58 + 2.34)} = 250.4 \text{ m}^3$$

The cut-fill ratio $= \frac{\text{Cuts}}{\text{Fills}} = \frac{2.58}{2.34} = 1.10$

Since the ratio is more than 1 it is acceptable.

Topographic Surveying

Topographic survey is essential to prepare a topographic map required for land levelling. It is a practice to determine the differences in elevation (vertical distances) between various points in the field. Surveying needs a chain (200 m), measuring rods (2-5 m), a measuring tap, a carpenter level, ranging poles (straight wooden rods about 2 m long), pegs and a dumpy level. A dumpy level (Fig. 6.2) consists of a telescope having an objective lens, an eye-piece lens with a cross hair in its diaphragm and a focussing screw. A spirit level rests on the telescope which in turn is attached to a levelling head having three levelling screws for levelling the telescope. Below this is a foot-plate which rests on tripod—head carrying tripod legs.

The instrument legs are spread 1.0–1.2 m and pushed firmly on the ground. The legs are adjusted to keep tripod-head levelled in both the directions (forward and backward). The telescope is turned until it comes over two opposite levelling screws and the bubble of the spirit level is brought to the centre by turning the screws. The telescope is then turned around with its stand over the third screw and the bubble is again brought to the centre by adjusting the screw. This exercise is repeated by turning the telescope each time by 180° and adjusting the bubble at the centre till it remains levelled in both the directions without disturbing the levelling screws.

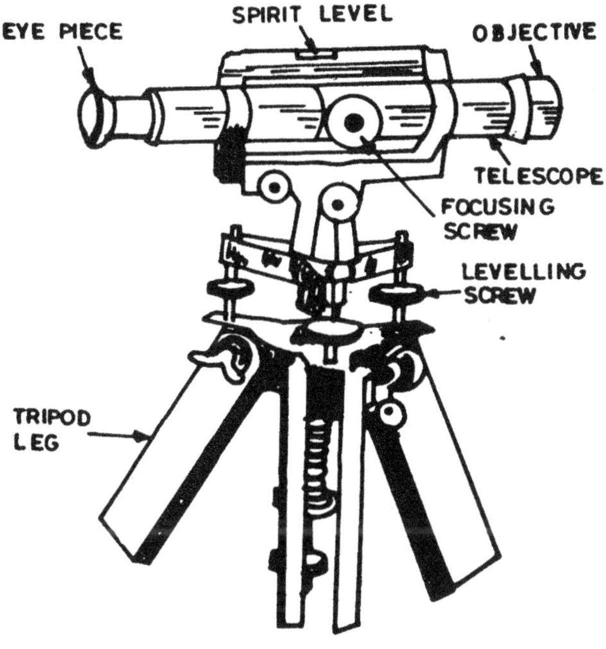

Fig. 6.2. A dumpy level

For measuring the elevation of a point, a level rod is installed vertically at the point and the telescope of the dumpy level (placed at a certain distance) is turned to bring the level rod in its field of vision. Then the eye piece is adjusted to focus it on the cross-hair. When properly focussed, the cross-hair stands out clear and distinct. The instrument is focussed on the rod by turning the focussing screw. When the instrument is in proper focus, the eye can be moved up and down in front of the eye piece and the cross-hair will not appear to move. The rod level is read at the intersection of the middle horizontal cross-hair with the rod. The level (ranging) rod is next moved to another grid point and the telescope is focussed again and the next reading is taken. In this way the entire field is measured and an elevation map of the topography is drawn.

Chapter — 7

Soil Conservation

Natural resources conservation and their efficient utilization for the benefit of mankind is the ultimate aim of watershed development. Soil erosion by water is one of the important sector of activities undertaken under this programme.

Soil Erosion by Water

Soil erosion by water is a major problem of the watershed areas. Water erosion is caused due to dispersive action and transporting power of water, particularly the rain water. The rain drops strike the ground with certain kinetic energy and splash soil particles from the land surface by their impact. These particles when come to water make suspension which gets entrained in runoff water. If there is no runoff, there will be no erosion. On the other hand, if there is no impact of rain drops there will be no erosion. In fact rain drops strike the ground with considerable impact and can splash over 60 cm high and 150 cm laterally on level surface. Usually, they have enough intensity to produce runoff over the land surface. Most of the rainfall energy gets dissipated in friction with the land surface and only 3 to 4 per cent of the energy of running water and 0.2 per cent of that of falling rain drops is expended in erosion.

The Running Water has Following Actions

Hydraulic Action: This is scouring up of soil particles due to hydraulic pressure of the water. The running water compresses the air present in the soil pores which in turn exerts pressure on the soil particles, resulting into their detachment. This action is more pronounced in a loose soil. Large particles of the soil together with rock fragments may thus dislodged.

Abrasion: The soil particles which are dislodged from the earth's surface run with the running water and scour the land surface causing more particles to be eroded away.

Attrition: The particles in the running water collide with each other resulting into further breakdown of the particles.

Solution: The soil particles get into solution with water due to chemical action. The dissolved particles carried away easily along with the running water.

Types of Water Erosion

The soil erosion by water is of three types:

(i) ***Sheet Erosion***: It is uniform removal of soil in thin layers like a sheet from gently slopy land resulting from overland flow occurring in thin layers. This idealized form of erosion rarely occurs. Rilling in fact takes place simultaneously with the first detachment and movement of soil particles.

(ii) ***Rill Erosion***: The beating action of rain drops combined with surface flow of rain water causes the initial microscopic rilling. The eroding and transporting power of sheet flow are functions of the depth and velocity of runoff for a given size, shape and density of soil particles. Rill erosion is the removal of soil by water from small, well defined channels when there is a concentration of overland flow. Conventionally, rill erosion occurs when these channels have become sufficiently large and stable to be readily seen. Rills are small enough to be easily removed by normal tillage operations. Although, rill erosion is often overlooked but it is the form of erosion in which most top soil loss occurs.

(iii) ***Gully Erosion***: The erosion having channels larger than rills and which can not be removed by tillage operation is called gully erosion. It is the advanced stage of rill erosion. Gullies are relatively permanent steep-sided water courses which experience ephemeral flow during rain storms. They are characterized by a head cut and various steps along their course and have relatively greater depth and smaller width and carry large sediment load. They occur on slopy lands.

Factors Affecting Soil Erosion

The factors which affect soil erosion by water are illustrated below:

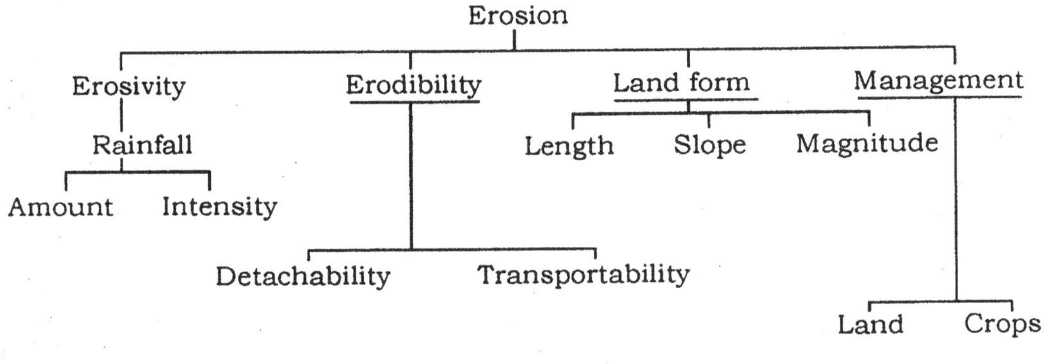

Erosivity: It is solely a property of rainfall and is closely related with the rainfall amount and intensity. Average soil loss per rain event increases with the intensity and amount of the rain storm.

Rainfall Intensity: In general, tropical rains are more erosive than temperate because of their higher intensities. The force responsible for detachment of soil particles is associated with the impact of individual rain drop. The question of whether it is due to momentum or kinetic energy (K.E.) of the drops, is still a matter of debate but more reports are in favour of K.E. of the rain drop. The K.E. depends both on size of rain drop and its terminal velocity and is found to be related with rainfall intensity (I) by the relation:

$$KE = 210.3 + 89 \log I \qquad (7.1)$$

where KE = kinetic energy (t/ha-cm) and I = rainfall intensity (cm/h).

Drop Size: Drop size, distribution and shape, all influence energy momentum and erosivity of a rain storm. The relationship between median rain drop size (D_{50} in mm) and rainfall intensity (I, cm/h) is given by the relation:

$$D_{50} = 2.64\ I^{0.82} \qquad (7.2)$$

This equation is valid for rain drops upto size 7.00 mm which is most common.

Terminal Velocity: The K.E. of a rain storm is related to the terminal velocity of the rain drop at the time of impact with soil by the relation:

$$KE = 1/2\ mv^2 \qquad (7.3)$$

where m = mass of rain drop and v = velocity of rain drop.

Wind Velocity: Wind accompanying a storm influence its K.E. and hence its erosive capacity. The rain accompanying wind has more K.E. than rain without wind.

Direction of Land Slope: The direction of land slope with respect to the direction of rain storm significantly alter the K.E. of the drop impact because of the angle at which it hits the ground. The maximum impact will be for normal (90°) incidence of the rain drops.

Erodibility: It is the resistance of the soil to both detachment and transport of soil particles and varies with soil texture, aggregate stability, shear strength, infiltration capacity and organic and chemical contents of the soil. The coarse sized particles are resistant to transport because of their more weight while finer particles are resistant to detachment because of their cohesiveness. The least resistant particles are silts and fine sands. Soils containing 40-60 per cent silt are more erodible while those containing 30-35 per cent clay particles combine with organic matter to form soil aggregate and hence less erodible. Non-expanding type clays provide more stable aggregates than the expanding type clays. The proportion of water stable aggregates less than 0.5 mm in a soil is a good index of erodibility. The erodibility increases with the increase in aggregates of size less than 0.5 mm in a soil. Sandy soils are more erodible than clayey soils as the aggregates of the sandy soils slake more readily and seal the soil surface. Divalent cations, like Ca^{++} is more helpful in aggregate stability than monovalent cations, like Na^{+}. Soils containing less than 3.5 per cent organic matter are unstable.

The silica-sesquioxide ratio is the most important index of soil erodibility. Erodibility decreases with the decrease of this ratio. Soils having this ratio less than 2 are more resistant to erosion.

Estimation of Soil Loss

The widely used method of soil loss estimation is use of the Universal Soil Loss Equation (USLE) given by Wischmeier and Smith (1960) which is expressed numerically as:

$$A = RKSLCP \quad (7.4)$$

where, A = average annual soil loss, t/ha

R = rainfall erosivity factor,

K = soil erodibility factor,

SL = slope-length factor,

C = cropping management factor which is ratio of soil loss for a given condition to that from cultivated continuous fallow, and

P = conservation practices factor which is ratio of soil loss for a given practice to that for up and down the slope farming.

This equation is based on observations recorded from the standard plots located at different places. It estimates soil loss from sheet or rill erosion under specified conditions.

***Rainfall Erosivity Factor* (*R*):** The rainfall factor is defined as one hundredth of the product of kinetic energy (KE) and maximum 30 minutes rainfall intensity (I_{30}) of a rainstorm. Annual total of each storm value of EI_{30} is a measure of R. It can be expressed as:

$$R = \sum (KE \times I_{30})/100 \quad \ldots (7.5)$$

The KE is related with rainfall intensity by the equation (7.1). Studies have shown that the median drop size of rainfall does not continue to increase with rainfall intensities more than 7.6 cm/h. Therefore, a limit of 7.6 cm/h be used in the computation of I_{30}.

Since rainfall intensity data are usually not available, an empirical expression for computing R was developed for West Africa:

$$R = (0.50 + 0.05)\ H \quad \ldots (7.6)$$

where H = rainfall amount, mm.

The relation was found appropriate to several locations in Ivory Coast, Upper Volta, Senegal, Niger and Tchad. The critical rainfall intensity was found to be more than 25 mm/h for tropical countries and 6-10 cm/h for temperate conditions.

Soil Erodibility Factor (K): This factor relates the rate at which different soils erode depending upon their inherent soil properties. Direct measurement of K is time consuming and costly and is feasible for only a few major soil types. Many attempts have been made to correlate K with sand, silty, clay, organic matter, aggregation index, antecedent soil moisture, structure etc. But the Table 7.1 given below which has been developed by the United States Soil Conservation Services, is commonly used.

Table 7.1: K value in relation to soil texture and organic matter.

Texture class	*Organic matter content, %*		
	≤ 0.5	2	4
Sand	0.05	0.03	0.02
Fine sand	0.16	0.14	0.10
Very fine sand	0.42	0.36	0.28
Loamy sand	0.12	0.10	0.08
Loamy fine sand	0.24	0.20	0.16
Loamy very fine sand	0.44	0.38	0.30
Sandy loam	0.27	0.24	0.19
Fine sandy loam	0.35	0.30	0.24
Very fine sandy loam	0.47	0.41	0.33
Loam	0.38	0.34	0.29
Silt loam	0.48	0.42	0.33
Silt	0.60	0.52	0.42
Clay loam	0.27	0.25	0.21
Sandy clay loam	0.28	0.25	0.21
Silty clay loam	0.37	0.32	0.26
Sandy clay	0.14	0.13	0.12
Silty clay	0.20	0.23	0.19
Clay	0.13	-	0.29

Topography Factor (L.S.): Erosion increases with the increase in slope steepness and slope length due to their respective increase in velocity and volume of surface runoff. Of the two characteristics of slope, degree is usually more important than length from the stand point of the severity of erosion. Experiments have shown that on slopes below 10 per cent, the amount of erosion, approximately, doubled as the degree of slope doubled. The product of slope length and slope gradient is called topographic factor and is given by

$$LS = \frac{(x)^m}{(22.13)}(0.065 + 0.045\ S + 0.0065\ S^2) \quad \text{... (7.7)}$$

where x = field slope length, m, S = slope gradient, per cent and m is constant whose value varies with S as follows:

m = 0.05 for S > 5%

m = 0.04 for S 3–5%

m = 0.03 for S 1–3%

m = 0.2 for S < 1%

Crop Management Factor (C): This term reflects combined effect of cover crop sequence, productivity level, length of growing season, tillage practices and residue management on erosion control. Evaluation of C factor is often difficult and is usually taken from standard tables. The values of C for different land conditions are given in Table 7.2.

Table 7.2. Value of C for different land conditions.

Land condition	*C value*
Fallow	1.0
Crop cover	0.4—0.3
Over-grazed grass	0.1
Thick grass	0.01
Forest (thick)/mulched	0.001
Pearl-millet	0.61
Pearl-millet followed by cowpea	0.39
Sorghum followed by pegeonpea	0.28
Rice	0.1—0.2
Palm tree	0.1—0.3

The value of C is higher at the early stage of crop growth and decreases with the increase in leaf area index. Vegetation on land surface reduces soil erosion. For adequate protection at least 70 per cent of the ground surface must be covered with vegetation. The value of C also depends on plant height.

Conservation Practices Factor (P)

The erosion control practices are contouring, strip cropping and terracing. The values of P for three major mechanical practices are given in Table 7.3.

The permissible limit of soil erosion may range from 25, 6 to 11 and 2 tons/ha/year, respectively, for mountainous land, deep fertile soils and thin highly erodible soils.

Table 7.3: Conservation practices factor (P)

Land slope, %	*Contouring*	*Strip cropping*	*Terracing*
1–2	0.60	0.30	0.12
3–8	0.50	0.25	0.10
9–12	0.60	0.30	0.12
13–16	0.70	0.35	0.14
17–20	0.80	0.40	0.16
21–25	0.90	0.45	0.18

Erosion Control

The aim of soil conservation is to obtain the maximum sustained level of production from the land, while maintaining soil loss below a threshold level which theoretically permits the natural rate of soil formation to keep pace with the rate of soil erosion. There may also be a need to reduce soil erosion to prevent nutrient loss, siltation of water bodies and pollution of water. Since, erosion is a natural process which can not be prevented but can be reduced to an acceptable level.

The strategy of soil conservation is based on covering the land surface to prevent it from raindrop impact, increasing infiltration capacity of the soil to reduce runoff, improving stability of soil aggregates to reduce splash and entraining of soil particles and increasing land surface roughness to reduce amount and velocity of runoff water. The practices adopted could be agronomical (biological), soil management and mechanical (engineering).

Agronomical Measures: The agronomic measures include covering soil surface by mulching, growing cover crops like cowpea, moong-bean, mothbean; grasses like sod, sudan, bhabhar (*Eulaliopsis binata*) fodders like alfalfa, berseem, trees (forest) like shisham (*Dalbergia sissoo*), kheir (*Acacia catechu*), babool (*Prosopis juliflora*).

Pasturing and Rotational Grazing; Strip Cropping and Multiple Cropping

Row crops (maize, oat) are least effective and therefore they must be combined with protective crops in a logical cropping pattern to reduce soil erosion. Legume-cereal, legume-fodder and legume-grass rotations are effective erosion control measures. Agronomic measures help to reduce kinetic energy of rain drops as well as of runoff water. The importance to mulch in erosion control is well known and is efiective for the periods when the crop itself does not adequately protect the soil. Mulches are more effective on land capacity classes II, III and IV. Covering soil surface with crop residue can reduce soil loss by 50 per cent or more depending upon the amount of rainfall and mulch, soil type and environment characteristics. Mulching material could be easily and cheaply available such as straw, stubble, cow dust, plastic film etc. The quantity of mulch may correspond to 1.5–2.0 tonnes of straw or 4.0–5.0 tonnes of farm yard manure per hectare.

Soil Management: The soil management practices include maintenance of soil fertility and building of soil structure. Highly fertile soil results in high crop yields and good plant cover and is therefore, resistant to soil erosion. These soils have a stable (usually granular) structure, which does not break down under cultivation, and a high infiltration capacity. Soil fertility can, thus, be regarded as the key to soil conservation.

Sands and coarse loamy sands have high infiltration rates and sand particle of size more than 0.3 mm are not easily eroded by running water or rain drop impact. Soils containing 40–60 per cent silt are easily erodible but containing more than 30 per cent clays are generally more coherent and form stable soil aggregates which are resistant to erosion. Soil aggregates of size more than 0.5 mm are less susceptible to water erosion, Clays are often clody soils and their rough surface store much water and are resistant to sheet and rill erosion. Soil structure can be improved by building organic matter in the soil (> 2%) by green manuring, suitable tillage practices and crop rotation. Application of corn straw at the rate of 1 kg m^{-2} per year can build up soil organic matter by about 0.017 per cent. Tillage provides a suitable seedbed for plant growth and helps to improve infiltration and reduce runoff. Stony soils and soils with a prismatic structure are less erodible. Soil measures help to reduce splashing of soil particles besides speed and amount of runoff water.

Mechanical Measures: Mechanical field practices are used to control the movement of water over the land surface. The practice depends on whether the objective is to reduce the velocity of runoff water or to increase surface water storage capacity or to safely dispose of the excess water. Mechanical measures are normally used in conjunction with agronomic measures. The various mechanical measures are contouring, strip cropping and terracing.

Contouring, Strip Cropping and Terracing (Engineering measures)

The important principles to be kept in view while planning these measures are

(i) Increasing the time of concentration and thereby allowing more runoff water to be absorbed and held in the soil profile.

(ii) Intercepting a long slope into several short ones so as to maintain less than a critical velocity for the runoff water.

(iii) Protection against damage owing to excessive runoff.

Contouring: It is a practice to carry out all farm operations like ploughing, seeding, inter-culturing on contours. By this practice soil loss from gently slopy land can be reduced up to 50 per cent. The technique is effective for low intensive rainfalls and land slopes up to 10 per cent. However, contouring along with strip cropping can take care of high intensity rainfalls. Contouring helps in reducing velocity of over land flow. If ridger cultivation is practised, the storage capacity of the furrow is materially increased, permitting storage of large volume of rain. When this practice is followed on steeper slopes or under conditions of high rainfall intensity and soil erodibility, there is an increased hazard of gullying because furrow breaks may release the stored water causing accumulative damage.

Contour Bunds: They are earth bunks, 1.5 to 2.0 m wide, thrown across the slope to act as a barrier to runoff, to form a water storage pool on their up slope side, and to break up a slope into segments shorter in length than is required to generate overland flow at critical velocity. They are suitable for slopes up to 6 per cent and in special cases up to 12 per cent. The vertical interval between two bunds can be estimated by the formula

$$VI = \frac{(2 + S)}{(4)} 0.3 \qquad \text{... (7.8)}$$

and

$$HI = \frac{VI \times 100}{S} \qquad \text{... (7.9)}$$

where VI = vertical interval (m), HI = horizontal interval and S = slope (%).

The bunds spaced at 10–20 m intervals are generally manually constructed. These bunds usually are of 0.8 m height (for depth of imponding 30 cm, depth of flow 30 cm and free board 20 cm), top width 0.5 m, bottom width 2.0 m and side slope 1:1. However, these dimensions can vary with the soil type. The bunds should be in a position to store the 24-hour rain storm expected in a 10-year return period. They are essentially meant for storing rain water *in situ* and are constructed in low rainfall areas (< 600 mm/annum). The depth of water to be imponded against the bund can be worked out by the formula.

$$F = \sqrt{VI. \frac{R}{500}} \qquad \text{... (7.10)}$$

where F depth of water to be imponded (m), VI = vertical interval (m) and R = maximum rain water on area basis to be stored, mm.

Graded Bunding: Graded bunds are laid along a pre-determined longitudinal grade (0.2 to 0.4 per cent depending on soil) instead of along the contours. They are constructed by constructing wide and relatively shallow channels across the slope and very near to the contour ridges and at critical interval to act primarily as drainage channels for inducing and regulating the excess runoff to drain at non-erosive velocity. It is adopted in areas receiving rainfall more than 800 mm/annum and on heavy soils. The vertical and horizontal spacing of the graded bunds can be estimated similar to those of contour bunds. However, extra spacing up to 25 per cent may be provided for highly permeable soils but it should be reduced by 15 per cent for slow permeable soils without any cropping programme. The terrace usually have a cross section of 1 m^2 and depth of 0.45 m. The grade should be changed at an interval of 100 m. The critical velocity of water flow in the channel should not be more than 0.5 and 0.6–0.75 ms^{-1} for sandy and clayey soils, respectively. They are suitable for land slopes up to 12 percent.

Strip Cropping: It is the practice of growing alternate strips of different crops in the same field. The strips are on contours in humid region and are of three types:

(i) contour strip cropping in which crops and their rotations are fixed on definite and suitable contours of land,
(ii) field strip cropping in which crop strips of uniform width are placed across the gentle slope of the land and
(iii) buffer strip cropping with strips of grass/legume crop laid out between contour strips of crops in the regular rotation and are placed at critical slopy areas of the field and may be regular or irregular in width.

The crops for strip cropping are alfalfa, cowpea, moong bean, potato etc. The width of strip cropping can be estimated by the relation:

$$W = 50.4 - 2.1\,S \qquad \text{... (7.11)}$$

where W = width (m) and S = slope (%).

Terracing: Terraces are earth embankment constructed across the slope to shorten slope length and to intercept surface runoff and convey it to a stable outlet at a nonerosive velocity. They thus, perform functions similar to contour bunds. They differ from them by being larger and designed to more stingent specifications. Terraces are of 3 types: diversion, retention and bench (Fig. 7.1). The diversion terraces are constructed to deliver the water to a suitable outlet. They, therefore, run at a slight gradient (1 : 250) to the contour. Diversion terraces are not suitable for slopes more than 12 per cent because

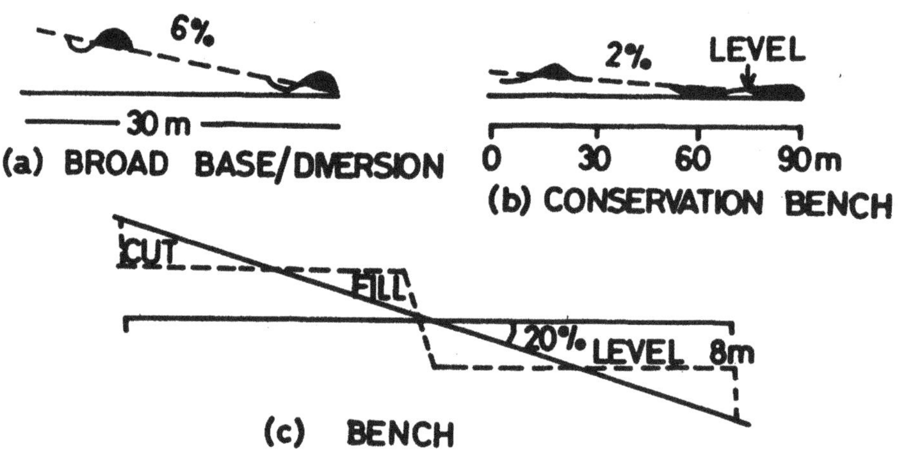

Fig. 7.1. Types of terraces

A View of Field Terracing for Soil and Rain Water Conservation

of high cost and the close spacing which would be required. Broad-based diversions usually have embankment and channel, each of width about 15 m, while, norrow-based diversions have these widths 3-4 m, and side slopes less than 15 per cent.

Retention terraces are used where it is necessary to conserve rain water *in situ*. They are levelled and designed to store runoff of a 24-hour duration in a 10-year cycle. They are suited for permeable and retentive soils on slopes of less than 8 per cent. They are broad-based channel and their function is to drain surface runoff or to absorb it. The horizontal distance (HD) and vertical distance (VD) may be determined by the formula.

$$HD = \frac{200}{S} + 10 \text{ and } VD = 2\frac{S}{10} \qquad \text{... (7.12)}$$

where S = slope (%).

Bench Terrace: Bench terraces consist of a series of alternate broad-based channels and small ridges and are employed where steep slopes up to 20 per cent need to be cultivated. In many situations, slopes up to 10-12 per cent is considered maximum. In Himalayan regions or Punjab and Himachal Pradesh, cultivation is extended up to 25 per cent while in southern states of Andhra Pradesh and Orissa agriculture is produced up to 10 per cent slope. They are constructed by making steps-like field along contours by half cutting and half filling and thereby slopy lands are converted into level fields. The side slope of channel or ridge is kept as flat as possible to facilitate modern farm machine operations. The specifications of broad-base terrace adopted in black soils are given below in Table 7.4.

Table 7.4: Specifications of broad base terrace in black soils

Top width (m)	*Base width (m)*	*Height (m)*	*Side slopes (H:V)*	*Cross section (m^2)*
0.3	4.85	0.45	5:1	1.15
0.3	6.40	0.80	5:1	1.98
0.3	5.60	0.45	5:1	1.35
0.3	7.60	0.60	6:1	2.34

Alignment of Bench Terraces: Alignment of the bench terraces should start from the ridge and progress towards the valley. The average land slopes of the area to be terraced should be determined by taking levels and then the specifications of the terrace are worked out. Contour lines may be worked out with the help of A-frame. Taking a contour line as the centre line, the terrace width may be marked on the ground. The alignment may then be examined with the considerations to the local conditions, that exist at the site such as depressions, sharp turns, field boundaries etc. and suitable adjustments made wherever necessary. Construction of the terraces may be started from the highest terrace and proceed downward. In order to maintain original soil fertility of the top soil, the terraces may be constructed from the bottom and after the construction of the first terrace, the top soil from the second terrace may be spread on the first terrace and the process continued for subsequent terraces. In bench terraced areas, suitable outlets should be provided to dispose off safely the runoff. This is possible by having suitable water-ways.

Terrace Cross-Section: The construction of bench terrace (Fig. 7.2) is such that the earthwork excavated from the upper half is deposited over the lower slope. The height of the embankment should be increased sufficiently to provide for shrinkage of soil. The entire terrace should be consolidated to make it stable.

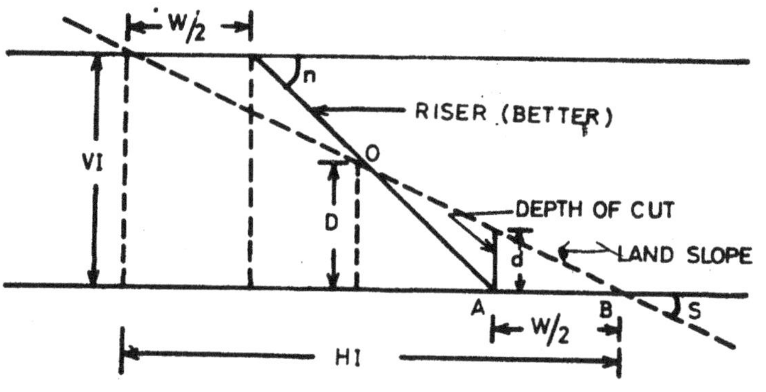

Fig. 7.2. A bench terrace

The earth-work per hectare of land (E_W) in terracing is computed as follows:

The earth work per unit length in digging the triangle

$$OAB = \frac{1}{2} \text{ base} \times \text{height} \qquad \text{(Fig. 7.2)}$$

$$= \frac{1}{2}\frac{(W)}{(2)} \times (D/2)$$

$$= \frac{W}{8}(D) = \frac{W.\ VI}{8}$$

∴ The earth work per hectare

$$= \frac{W}{8}(VI) \times L/L.\ HI, \quad \text{where } L = \text{total length of terrace/ha}$$

$$= \frac{W}{8}(S \times HI)\ L/L.\ HI$$

$$= \frac{W}{8} \cdot S \qquad \dots (7.13)$$

Area available for cultivation

$$= 100\,(10 - nS) \qquad \text{... (7.14)}$$

Area lost due to benching

$$= n \times D\,(L)$$

$$= nS \times HI \times L \quad (\because\; D = VI = S.\,HI)$$

$\therefore$ Area lost per ha in benching

$$= nS \times HI \times L/HI.\,L$$

$$= nS$$

$$= 100\,nS \text{ (if S is in percentage)} \qquad \text{... (7.15)}$$

$$\text{Better area to be grassed} = 100\sqrt{1 + n^2} \qquad \text{... (7.16)}$$

$$\text{Number of terrace outlets/ha} = L/2K \qquad \text{... (7.17)}$$

where K = critical length of terrace. It is that length of a terrace at which the rainfall runoff attains critical velocity for soil erosion.

Construction of Hill Terraces

The upper reaches of a watershed usually consist of hilly areas. These areas have an undulating topography and are susceptible to soil erosion. In areas with steep slopes and subjected to soil erosion, vegetation do not get established. Due to lack of vegetation cover, soil erosion is accelerated transporting large quantities of silt into streams below. The uncontrolled runoff from the sloping areas caused extensive damage to the down stream agricultural lands and other areas. When hill slopes are put under cultivation erosion becomes more severe due to the soil tillage required for agriculture. It is, therefore, necessary to treat the steep hill slopes under soil erosion with appropriate soil conservation measures so that these are covered with vegetation and the surface runoff from these areas is controlled. Contour trenching and afforestation, bench terracing and stone terracing are measures suitable for such areas.

Stone Terracing of Hill Slopes

Steep slopes are protected by constructing stone walls across the slope in areas with plenty of easily available stones. These stone walls called stone terraces, are small embankments constructed, with stones. The recommended specifications for stone terraces are given in Table (7.5). The cross section of the terrace may be 0.5 m^2 depending upon the depth of foundation. The foundation could be 15 cm for slopes less than 10 per cent and 30 cm for slopes more than 10 per cent.

Table 7.5. Specifications of stone terraces.

Slope (%)	*Horizontal distance (m)*
6	30
7	25–30
8	20–25
9	18–20
10	13–18
12	10–13
15	7–10

For constructing stone terraces, a shallow trench of one metre width and 15 to 30 cm depth (as the case may be) dug at the lower side of the base line and the spoil is spread towards upper side of the line to form a narrow ridge so that the percolation of runoff from field through the stone terrace is avoided. Stones should be packed into the foundation and above the ground level to form terrace. The stones should be interlocked atleast 7.5 cm in the joint and sharp edges should be exposed outside so that the proper shape of the bund is obtained. Alignment of stone terraces can be done with the help of Dumpy level or Abney level. The horizontal grade of 0.2-0.3 per cent may be provided towards outlet of the terrace.

***Contour Trenching*:** Contour trenching (Fig. 7.3) implies excavating trenches along the contour or along a uniform level. Contour trenches are used both on hill slopes as well as on degraded and bare wastelands for soil and moisture conservation and afforestation purposes. These trenches break the slope lengths, reduce the velocity of surface runoff and consequently retard

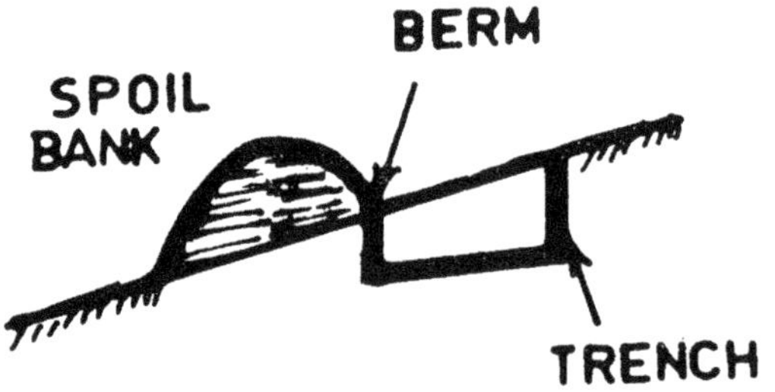

Fig. 7.3. Cross section of a contour trench

its scouring action and carrying capacity. The water retained in the trenches helps in conserving the moisture and provides advantageous sites for sowing and planting. Contour trenches can be used for any slope, rainfall and soil type, however, they are found economical for slopes less than 20 per cent. Contour trenches may be continuous or interrupted (Fig. 7.4) which may be in line or in staggered manner. Continuous trenches are used for moisture conservation in low rainfall areas and required careful layout. Interrupted trenches are used in high rainfall areas. The size of trenches depend upon the soil depth available at the site. Normally trenches of sizes between 1 to 2.5 m^2

A View of Contour Trenches

in cross-section with 0.3 m base, 0.3 m top, side slopes of 1:1 and trepezoidal in shape are practiced. However, the exact size and number of trenches can be calculated from the knowledge of rainfall required to be retained in an unit area. Preferably shorter lengths, 3 to 7 m are adopted for convenience of layout and construction. For areas where pasture is to be developed. long trenches up to 300 m may be laidout.

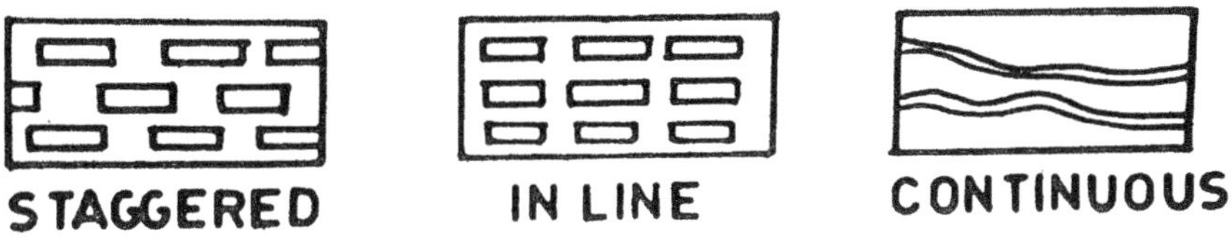

Fig. 7.4. A layout of contour trench.

The alignment of the trenches (i.e., marking the position of trenches on the ground should start from the ridge. After fixing the necessary horizontal interval, contour lines can be laid out with the help of a levelling instrument. A simple device known as the A-frame level (Fig. 7.5) is commonly used. To use this leveller, a peg is driven at the starting point and one end of this frame is placed against this peg and another end is moved up and down on the ground till the bubble of spirit level comes to the centre. By repeating this process, a line along the contour is laid passing through the 1, 2, 3 etc. pegs. After the contour line is stalked, the trenches may be marked with the help of a measuring tape. Similarly, the alignment of the subsequent contours may be done, fixing the requisite horizontal interval between the contours and the position of the trenches marked on them. Construction of the trenches should be started from the ridge and progessively extended towards the valley. In no case, the trenches in the down stream areas should be kept towards the trench as it can be used for refilling, if necessary, or otherwise it can be put on that portion of the spoil-bank where planting is to be done. Boulders and gravels from excavation should be stalked on the lower side of the trench to serve as a toe to the spoil-bank. Trenches should be dug during winter and spring so

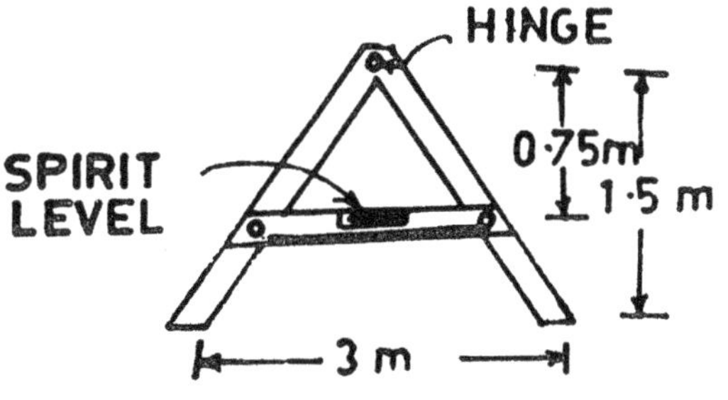

Fig. 7.5. A trench laying frame

that one site is ready for planting during the following monsoon season. Protection of the trenched area from animal and human interference is essential till it is vegetated to the required extent.

Waterways: The purpose of waterways in a conservation system is to carry out runoff at a non-erosive velocity to a suitable disposal point. Therefore, they should be in a position to carry the peak runoff from a storm with a 10-year return period. The waterways could be diversion channels, terrace channels and grass waterways. Diversions are placed up-slope of areas of a farm land to intercept water running off the slope. Terrace channels collect runoff from the inter-terraced area and convey it across the slope to a grass waterways. Grass waterways are designed to transport the runoff to a natural drainage system. They are located in natural depressions wherever possible and are recommended up to 20 per cent land slope. For slope more than 20 per cent stone lining should be done in the channel to protect erosion. The grasses commonly used are anjan (*Cenchrus ciliaris*), bermuda (*Cynodon doctylon*), kentucky blue (*Poa pratensis*), napier (*Pennisetum purpurecum*), etc.

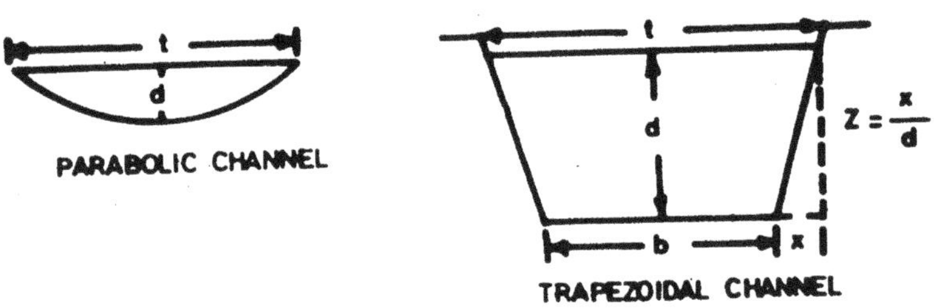

Fig. 7.6. A parabolic and trapezoidal channel

The grass waterways are designed on the principle of open channel hydraulics using the Mannings equation (7.24). In soil conservation work, these channels are designed to convey the peak runoff (cm/d) expected with a 10-year return period without causing scour or fill in the channel. Usually, trapezoidal or parabolic shaped channel (Fig. 7.6) are designed. The area of cross section (A), wetted perimeter (P) and top width (t) to trapezoidal section of a channel can be obtained by the expressions:

$$A = bd + Zd^2 \quad \text{... (7.18)}$$

$$P = b + 2d\sqrt{Z^2 + 1} \quad \text{... (7.19)}$$

$$t = b + 2Zd, \text{ and} \quad \text{... (7.20)}$$

$$R = A/P \quad \text{... (7.21)}$$

For parabolic channel

$$A = 2/3 \text{ td and } P = t + \frac{8}{3} d^2/t \quad \text{... (7.22)}$$

where b = bottom width, d = depth of channel, Z = side slope, R = hydraulic radius and t = top width. The wetted perimeter is the length of a channel cross section which gets wetted when the water flows in the channel to a depth d. The sides of earth channels are usually built steep so that the earth can withstand deformation when wet. The various values of Z and safe velocity for different soils and values of n for different land conditions are given in Table (7.6).

Table 7.6: Values of Z, safe velocity and n for different soil and land conditions

Soil texture	*Z (H/V)*	*Safe velocity* ms^{-1}	*Land condition*	*n*
Sandy and loamy sand	3 : 1	0.75	Earth	0.022
Sandy loam	2 : 1	0.90	Rocky	0.040
Loam and silt loam	1.5 : 1	1.00	Weedy	0.035
Clay	1 : 1	1.25	Short grass	0.045
Stiff clay	0.5 : 1	1.50	Medium grass	0.055
Rock	0.25 : 1	1.80	Long grass	0.095

For the most effective cross section of the trapezoidal channel the bottom width (b) and its depth (d) are related by the expression:

$$b = 2d \tan\frac{\theta}{2} \quad \ldots\ldots \qquad \ldots (7.23)$$

where, θ is the side slope of the channel. The peak rate of discharge is calculated by using the equation (4.7). The velocity of flow of water in the channel is calculated by the Manning equation:

$$V = R^{2/3}\ S^{1/2}/n \qquad \ldots (7.24)$$

where, V = velocity (m/s), R = hydraulic radius and S = bed slope of the channel.

The safe velocity of water in the channel varies with soil texture. For vegetative channel, the value of velocity could be taken as 1.20 ms^{-1}. For pipe (clay/concrete) the value of n is 0.015, 0.017 and 0.020, respectively, for pipe diameter of 7.5-20, 25-30 cm and 30-45 cm. For PVC corrugated pipes the value of n is 0.055.

Example: Rainfall run-off from a 20 m high hill at a slope of 45° from the horizontal is causing a waterlogging problem in the adjacent land with 1000 m length along the hillside. The peak intensity of rainfall is 70 mm h^{-1} Design a ditch for the safe disposal of the run off in a grassed loamy soil.

Solution: The runoff from the hill is given by equation (4.7)

$$Q_P \frac{CIA}{36} = \frac{0.82 \times 70 \times 20 \times 100}{36 \times 1000 \times 100} \qquad (C = 0.82 \text{ from Table 4.1})$$

$$= 0.32\ m^3s^{-1}$$

$$V = R^{2/3}\, S^{1/2}/n \text{ and } Q = AV = \frac{R^{2/3}\, S^{1/2}}{n} \cdot A.$$

For the loamy soil, side slope should be 1.5 : 1, safe velocity for grassed loamy soil should be 1.0 ms^{-1}, and $n = 0.045$ and $S = 0.01$

$$\therefore \qquad \tan\theta = \frac{1}{1.5} \text{ or } \theta = 33.7 \text{ and } Z = 1.5$$

and $$b = 2d \tan\frac{\theta}{2} = 0.6\ d$$

or $$R = A/P = \frac{bd + 2d^2}{b + 2d\sqrt{Z^2+1}} = \frac{0.6\ d^2 + 1.5\ d^2}{0.6\ d + 2d\,(1.8)} = 0.5d$$

or $$Q_P = 0.32 = (0.5d)^{2/3}.\ (0.01)^{1/2}.\ (2.1)\ d^2/0.04$$

and $$A = bd + Zd^2 = 2.1\ d^2$$

or $$0.128 = (0.5)^{2/3}\ d^{2/3}\ (2.1)\ d^2$$

or $$d^{8/3} = 0.0967$$

or $$d = 0.42 \text{ m and } b = 0.6\ d = 0.42 \times 0.6 = 0.25 \text{ m}$$

The corresponding $V = Q_P/A = \frac{0.32}{0.37} = 0.86\ ms^{-1}$ which is safe.

For $d = 0.40$, $V = 0.95$ which is also safe. Therefore, the channel of depth 0.40 m, bottom width 0.24 and top width 0.40 + 0.15 = 0.55 m should be dug across the hill length just at its foot to prevent the possible runoff to the land by diverting the water (runoff) through its channel.

Technology for Gully Control

The technology for gully (ravine) control consists of treating (a) level and marginal land and (b) gully itself. For level and marginal land diversion bunds may be constructed at a distance from the gully rim equal to twice the depth of gully and the head cross-section of the bund may be kept as 2.75 m^2. For lands up to 4 per cent slope, contour/graded bunds (1.0 m^2 cross section area) can be constructed at 0.5 to 1.0 m vertical interval. Drainage may be affected where drainage problems are anticipated. Channel of suitable capacity may be provided for safe disposal of excess runoff water.

For gullies, where no runoff is expected from top, earthen gully plugs of 1.1 m^2 cross section (22.5 cm below top level) and spaced at 45-60 cm horizontal interval are suitable. For gullies in which excess run off from the top is expected, an earthen gully plug of 2.25 m^2 cross section with a pipe outlet has to be provided. A pipe of diameter 15 cm of ACC is suitable to take care 0.05–0.07 cumecs discharge. A composite check dam of earth and brick masonry (spillways portion) would be required for catchment more than 1.6 ha and discharge more than 0.07 cumec, through big gullies. They are located at the confluence of a big compound gully at 1.2 m vertical interval on 1% slope of the gully bed. An apron is usually required to protect the structure from action of the runoff as it discharged from the spillways. Several low check dams are more desirable than one large dam of equivalent height. Low dams are less likely to fail and more easy to maintain than high one. The height of temporary dam should not exceed 0.6 m.

Depending upon the size of the gully, the following measures may be adopted:

(i) Small gullies (< 3.0 m deep) can be reclaimed by clearing, minor levelling and construction of a series of contour bunds at a vertical interval of 1.0 m. Pipe outlets are to be provided to drain excess runoff water. if any, from the treated area. At the end of small gully system, small earthen dam with a masonry spillway is provided to take excess runoff safely.

(ii) Medium gullies (3–9 m deep) can also be reclaimed by clearing, levelling the gully and construction of a series of composite earthen and brick masonry check dams at a vertical interval of 1.2 m with an outlet slope of 0.5-0.75 per cent between two consecutive structures and bench terracing the side slope (8-15%).

(iii) Deep gullies (> 9m) are reclaimed by afforestation with local tree species and constructing gully plugs in the bed of the main and branch gullies. Gully beds and sides are stabilized by planting grass species like *Dicanthium annulatum*. Gully plugging is done with small earthen check dams and live hedges.

Grass is established both on the bed and the sides either by seeding or by sodding. Sod (grass) flumes are used to control overfalls at the head of the small gullies. The head of the gully is shaped to a gentle slope not exceeding 6 : 1 and grass is established by sodding. Enough width is provided so that the flow velocities do not damage the grassed surfaces. In order to stabilize the toe of the flume, a wall of stones/bricks could be constructed.

Temporary structures viz., (i) brushwood dams, (ii) loose rock dams (iii) rock filled dams, and (iv) woven wire dams, depending on easy availability of the material and need, are provided to retard the flow of water and reduce the channel erosion. In the brushwood dams, locally available wood logs (10-15 cm diameter) are used while stones are used in rock dams and wire mesh is used to hold the stones in place in the woven wire dam.

Permanent protection of bunds and embankments can be had by facing them with a resistant material such as concrete. Still better, however, is to construct retaining walls with gabion structures. The gabion structure is a rectangular steel wire mesh basket packed tightly with stones—12.5 to 20.0 cm diameter. Gabions are supplied flat and then folded into rectangular shape on the site of installation. They are placed in position and anchored before being filled with stones. The gabion is filled to one-third of its depth and after which two connecting wires are inserted on further filling. The wires are repeated when the basket is 2/3 full. The gabion is slightly overfilled to allow for settlement. The hanged lid is closed and wired to the sides. The simplest structure consists of one tier of gabion, 1 metre high. A second tier can be positioned on top of the first set. The addition of further tiers however, requires structures against overturning.

Permanent structures are constructed when the benefits from such structures are justifiable compared to the cost of construction. The permanent structures protect the gullies from further development and at the same time help in storage of water. For the purpose of gully control, three types of structures are adopted viz; (i) chut spillway, (ii) drop spillway and (iii) drop-inlet or pipe spillway (Fig. 7.7). Chute spillways are used at the gully head to convey the water safely to the gully bed. The drop spillways are used along the gully bed to act as barriers so that gully bed is not eroded below the crest level of the structure. The drop inlet spillways are used at appropriate locations in the gully for storage of water.

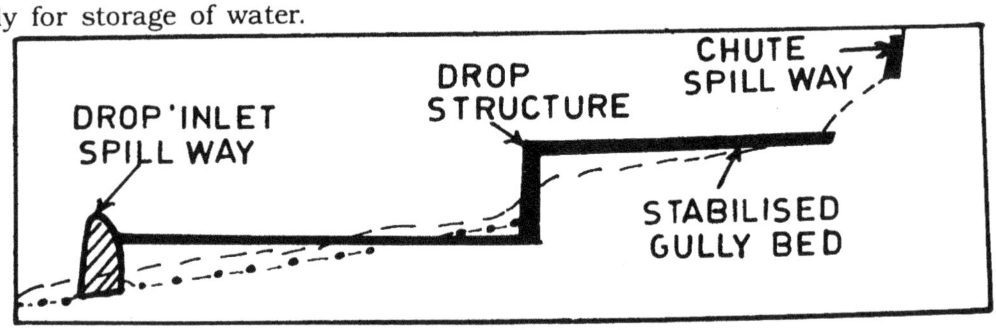

Fig. 7.7. Layout of a gully control structure

In areas where a considerable amount of silt is carried by the inflowing water, preventive measures to reduce silt deposited in the pond are essential, because removal of silt after it is deposited, is not possible to any appreciable extent. The preventive measures include:

(i) Converting slopes into vertical drops by terracing,
(ii) Stabilizing soil by turfing and by plant growth and
(iii) Constructing check dams.

A grassed filter strip in 3 meter length should be provided in the water way immediately above the pond. The perennial grasses used should create a high roughness value so that flow through the grass will be reduced in velocity to a point of non-erodibility.

Drop Structure: It is one of the most commonly used erosion control structure and consists of (1) head wall and head wall extension (2) side walls (3) wing walls (4) apron (5) longitudinal sill (6) end sill and cut off walls (Fig. 7.8). The length of the weir, depth of flow and dimensions of the structure are designed to provide a structure as such that maximum energy dissipation occurs. The crest length is calculated using the relation:

A Permanent Structure to Protect the Gully and Conserve Runoff

$$Q = 1.66\, Lh^{3/2} \qquad \dots 7.25$$

where Q is discharge (cumecs), L is length of weir (m) and h is head of flow (m). A free board of 15-20 cm may be provided on h. The critical depth (d) is worked out using the equation;

$$d = 3\sqrt{Q^2\, L^2 g} \qquad \dots 7.26$$

where g is the acceleration due to gravity.

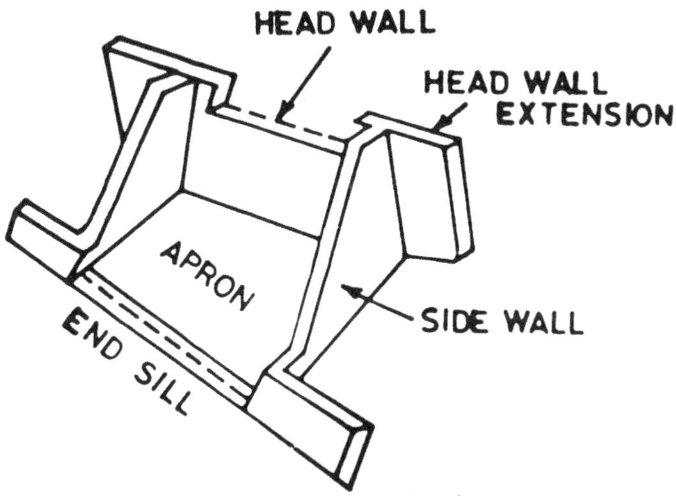

Fig. 7.8. A drop structure

The critical depth is useful to determine other dimensions of the structure. The objective of the drop structure is to dissipate the energy of the fall of water to an extent it does not cause damage down stream of the structure.

Surface Drainage: Surface drainage of excess water is done by open ditches. There are two kinds of ditches (i) field ditches and (ii) collection ditches. Field ditches are small ditches that collect water from the fields to be delivered to collection ditches that are big in size. The field ditches are shallow and as such they do not provide hindrance to normal field operations by farm machineries. They could be constructed manually or are properly spaced dead furrows constructed during ploughing operation. They are spaced 15-45 m apart, laid parallel to each other and 30-60 cm deep. The ploughing operation is done parallel to the ditches. The field ditches have bed slope almost equal to land slope and may range from 0.2–0.5 per cent and side slope of 3 : 1 or less depending upon the soil type.

Collection ditches may be V shaped or trapezoidal and can be constructed manually. V-shaped ditch is easier to be constructed by a grader. For design of a size, design criteria as discussed earlier in this chapter may be used. In the absence of any criterion, 10 cm of water removal in 24 hours may be taken as criterion. The side slope may be kept about 2 : 1 to 1.5 : 1 for sandy loam and clay loam soils. The velocity of water flow in the channel should range 75–90 cm/s. For stiff slopes, drop structures may be provided to reduce the velocity.

Problems

1. On a 20 per cent hill slope bench terraces are to be constructed at a vertical interval of 2 m; keeping cut equals to fill. Calculate (1) length of terrace per hectare; (2) earth work and (3) area lost for vertical cut. The terrace cuts are vertical.

Solution: It can be known from figure 7.2 that slope = arc/radius or S = 2d/w; where S = slope of the hill; d = depth of cut and W = width of terrace.

$\therefore$ W = 2 × 1/0.2 = 10 metres.

(1) Length of terrace per hectare = 10000/10 = 1000 m.

(2) Earth work per hectare = W.S/8 = 100.10.20/8 = 2500 m. [from equation 7.13].

(3) Since the terrace cut is vertical no area is lost in terracing.

2. Compute dimensions of a contour bund on a land slope 2%, vertical interval 1 m and maximum rainfall to be stored 50 mm.

Solution: From the equation 7.10 it is known that:

$F = \sqrt{VI.\ R/500} = \sqrt{1 \times 50/500} = 0.32$ m.

Allowing free board of 0.15 m; the height of bund

= 0.32 + 0.15 = 0.47 = 0.50. (say).

Top width = 0.3 m (assumed)

Side slope = 2:1 (assumed).

Bottom width = 2.3 m (calculated).

Area of cross section = [b + t]/2d = [2.3 + 2.3]/0.5/2 = 0.65 m^2.

Length of bund/ha = 100s/VI = 100 × 2/1 = 200 m.

Volume of earth work per ha = 100s × Area of cross section/VI

= 200 × 0.65 = 130 m.

3. Determine the bottom width, top width and total depth for a trapezoidal ditch to carry a flow of 0.5 m^3/s from a sandy loam soil.

Solution: For sandy loam soil the safe velocity of water in the channel is 0.9 m/s (from Table 7.6) and the slope of the channel should be 2 : 1.

$\therefore$ tan θ = 1/2 or θ = 26.6 and b = 2d tan θ/2 (from equation 7.23) = 2d tan 13.3 = 0.47d and

$R = bd + zd^2/b + 2d\sqrt{z^2+1} = 0.47d + 2d^2/(0.47 + 4.47d) = 0.5d.$

From the Mannings formula (equation 7.24), we have $V = R^{2/3} S^{1/2}/n$ and $Q = A.V = R^{2/3} S^{1/2}/n.\ (bd + zd^2)$ or $0.5 = (0.5d)^{3/2}\ (0.01)^{1/2}\ (2.5d)/0.04$ (S = 0.01)

assumed and $n = 0.04$; table 7.6). or d = 0.45 and corresponding V = 0.78 which is less than 0.90 m/s (safe velocity). The bottom width, top width and depth for the channel are b = 0.47d = 0.47 × 0.45 = 0.22, 2.1 and 0.46 m respectively. After adding a free board of 0.14 m to the depth, the total depth will be 0.60 m.

4. Determine the soil loss in tons/hectare for a field having R = 300, K = 0.5 (t/ha), S = 10%, L = 90 m and C = 0.2. The field is contoured (P = 0.6). By what measures the soil loss can be reduced to permissible limit of 7.5 t/h.

Solution: The soil loss equation is

A = RKSLCP (equation 7.4).

In order to find out the SL factor, the equation is

$$SL = (x/22.13)^m \{0.065 + 0.045\, S + 0.0065\, S^2\} = 2.356 \text{ \{for m = 0.5\}}.$$

$$\therefore\ A = (300)\,(0.15)\,(2.356)\,(0.2)\,(0.6) = 12.72 \text{ t/ha}.$$

The soil loss can be reduced by reducing C by using the land for forest (C = 0.2). Another way is to reduce P (by having strip cropping) from 0.6 to a low value for example to 0.5 by strip cropping in the field of maximum strip width 50 m.

Chapter — 8

Dryfarming Technology

Adoption of modern dryfarming technology ensures judicious management of land and water resources for the benefit of the watershed community and conservation of natural resources. Productive use of the most limiting resource water in dryland agriculture is possible by selection of short duration, low water requiring crops and their varieties and growing them by adopting the most appropriate agro-technique developed for the area. Comprehensive research carried out by several research organisations has helped to know the most efficient crops and their varieties for dryland agriculture.

There are various approaches to the selection of crops and varieties. Land use capability is the ideal concept but is rarely followed in dryland areas. It is moisture storage capacity of the soil and water availability that govern land use capability in dryland areas in addition to factors such as topography, erosion hazard, soil fertility etc. Moisture storage capacity depends on the depth, texture and structure of the soil.

Water Requirement of Crops

Information on total water requirement of crops, trees, shrubs and grasses is vital for selecting a suitable crop and cropping system. This information is usually available with the nearest Agricultural Experimental Station. It can also be worked out experimentally. For North-West dryland areas of India, the water requirement of different crops are reported in Table 8.1. In case the information about water requirement of a crop is not available, it could be estimated from the knowledge of Open Pan Evaporation (Ep) data as measured by various meteorological observatories. By multiplying Ep of the nearest observatory with a pan coefficient Kp (Table 8.2), the water requirement of the crop as potential evapotranspiration (ETo) can be worked out. The actual amount of water used by the crop that is water requirement (ETc) is related to the reference crop ETo by a crop coefficient K such that ETc = KETo.

Table 8.1: Water requirement of some crops in dryland areas of N-W India

Crop	*Water requirement (mm)*	*Crop*	*Water requirement (mm)*
Blackgram/cowpea (summer)	250	Sorghum	400
Castor	400	Soybean	500
Clusterbean	315	Sunflower	500
Cotton	600	Tobacco	500
Groundnut	450	Barley (Winter)	300
Maize	450	Chickpea	250
Moth bean	225	Lentil	280
Mung bean	240	Mustard	200
Pearlmillet	275	Rocket plant	150
Pigeon pea	450	Safflower	450
Sesamum	350	Wheat	350

The K value depends on crop characteristics, its sowing time, rate of development, growth stage and length of growing season besides climatic conditions particularly wind and humidity. The graph sowing variation in crop coefficient with growth stages of a crop is sown in Figure 8.1. The specific values of K for 3 and 4 growth stages are given in Table 8.3 while for 1 and 2 stages can be calculated from the graph matching with K values at 3 and 4 stages.

Table 8.2: Pan coefficient (Kp) for American Class—A type evaporation pan

Wind	*Cropped Area*			*Dry Fallow Area*		
	Relative humidity %					
	< 40	40–70	> 70	< 40	40–70	> 70
Light (< 2 m/s)	0.65	0.75	0.85	0.60	0.70	0.80
Moderate (2-5 m/s)	0.60	0.70	0.75	0.55	0.65	0.80
Strong (5-8 m/s)	0.55	0.60	0.70	0.50	0.50	0.65
Very strong (> 8 m/s)	0.50	0.55	0.60	0.40	0.50	0.55

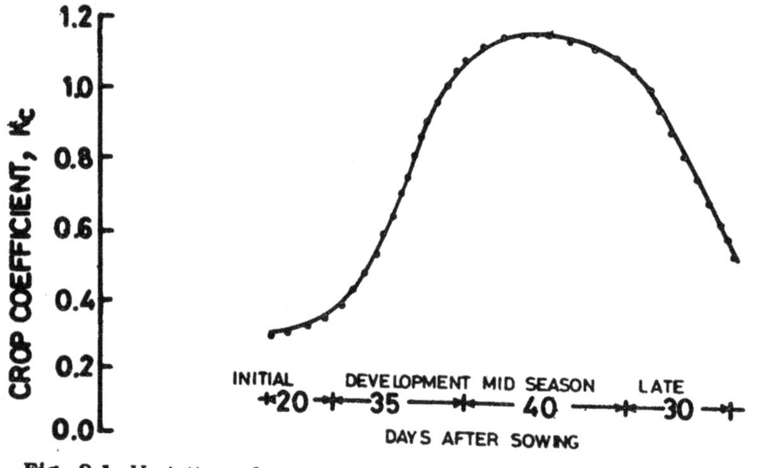

Fig. 8.1. Variation of crop coefficient with stages of crop growth

Table 8.3: Crop coefficient for different stages of crop growth and climatic conditions

Crop	*Relative humidity %*			
	High > 70%	*Wind, m/s*	*Low< 20%*	
	0–5	5–8	0–5	5–8
Barley	1.05/0.25	1.10/0.25	1.15/0.20	1.20/2.20*
Beans	1.05/0.30	1.10/0.30	1.15/0.25	1.20/0.25
Carrot	1.00/0.70	1.05/0.75	1.10/0.80	1.15/0.85
Corn	1.05/0.55	1.10/0.55	1.15/0.90	1.20/0.60
Groundnut	0.95/0.55	1.00/0.55	1.05/0.60	1.10/0.60
Lentil	1.05/0.30	1.10/0.30	1.15/0.30	1.20/0.25
Millet	1.00/0.30	1.05/0.30	1.10/0.25	1.15/0.25
Oat	1.05/0.25	1.10/0.25	1.15/0.20	1.10/0.20
Onion	0.95/0.75	0.95/0.75	1.15/0.80	1.20/0.80
Pea	1.05/0.95	1.05/0.95	1.10/1.00	1.20/1.20
Potato	1.05/0.75	1.15/0.75	1.15/0.75	1.20/0.75
Sunflower	1.05/0.25	1.05/0.25	1.15/0.20	1.20/0.20
Sorghum	1.00/0.50	1.05/0.50	1.10/0.55	1.10/0.55
Soybean	1.00/0.45	1.05/0.45	1.10/0.45	1.10/0.45
Sugarbeat	1.05/0.90	1.10/0.95	1.10/0.95	1.15/1.00
Sunflower	1.05/0.40	1.10/0.40	1.15/0.35	1.20/0.35
Tomato	1.05/0.60	1.10/0.60	1.20/0.65	1.25/0.65
Watermelon	0.95/0.65	1.05/0.65	1.00/0.75	1.05/0.75
Wheat	1.05/0.25	1.10/0.25	1.15/0.20	1.20/0.20

* The lower figures are for stage 4 while the upper figures are for stage 3 of the crop.

Efficient Crops and Cropping Systems

Productive use of water in dryland is possible by selecting short duration, drought tolerant, low water requiring and high yielding crops and their varieties and growing them by adopting the most appropriate and accurate agro- techniques. Comprehensive research on identification of efficient crops for various agroclimatic situations has suggested the following most efficient crops suitable for different soil types and climatic conditions.

Soil type	*Climate*	*Crops*	
		Summer	*Winter*
Alluvial	Arid	Pearlmillet, Mungbean	–
	Semi-arid	Pearlmillet, Mungbean, Cowpea, Blackgram, Sesamum, Clusterbean Pigeonpea, Maize	Chickpea, Mustard
	Sub-humid	Cotton, Maize, Pigeonpea Sorghum	Barley, Chickpea, Lentil, Wheat

Black	Arid	Sorghum	-
	Semi-arid	Sorghum, Caster Pigeonpea, Mungbean, Soybean	Sorghum, Chickpea, Safflower
	Sub-humid	-do-	-do- Pea, Wheat
Desertic	Arid	Pearlmillet	-
	Semi-arid	Pearlmillet, Mungbean Blackgram	Chickpea
Red	Arid	Finger millet	-
	Semi-arid	Mungbean, Groundnut Pigeonpea, Soybean, Sorghum, Finger millet	Sorghum
	Sub-humid	Maize, Rice	Chickpea, Linseed, Sunflower Wheat
Submountainous	Semi-arid	Maize	Wheat
	Sub-humid	-do-	-do-

The choice of variety changes frequently due to continuous release of new crop varieties. The selection of variety should be made from the latest released varieties.

Cropping System

A cropping system refers to a combination of crops in time and space. Combination of time refers to the growing of different crops in different seasons of a year while the space combination refers to situations when two or more crops are inter-planted as inter-cropping. Cropping system approach aims at increasing the productivity per unit area per unit time by efficient utilization of natural resources such as land, water and solar radiation besides external inputs such as seed, fertilizer, pesticides, labour, credit and power. It also needs to consider the market value and the demand of farm produce. For different agro-climatic regions and soil types some of the important cropping systems identified are as under.

Soil type	*Climate*	*Cropping system*	
		Summer	*Winter*
Alluvial	Arid	Pearlmillet/Mungbean/ Blackgram/Clusterbean Fallow	- Fallow - Chickpea
	Semi-arid	Pearlmillet	- Chickpea
		Mungbean/Blackgram/Clusterbean	- Mustard - Fallow
		Fallow	- Chickpea/Mustard
	Sub-humid	Maize	- Wheat/Lentil/ Toria/Sunflower
		Arhar	- Fallow

Black	Arid	Fallow	- Sorghum
		Sorghum	- Fallow
	Semi-arid	Mungbean	- Chickpea/ Sunflower
		Pigeonpea	- Fallow
		Sorghum/Soybean	- Fallow
	Sub-humid	Maize/Sorghum	- Sorghum/ Chickpea Sunflower/Pea
		Soybean	- Wheat
Desertic	Arid	Pearlmillet	- Fallow
	Semi-arid	Pearlmillet/Mungbean/ Blackgram/Clusterbean	- Fallow/ - Chickpea/Mustard
		Fallow	- Chickpea
Red	Arid	Millet	- Fallow
	Semi-arid	Mungbean/Soybean/ Groundnut/Pigeon Pea	- Sorghum
	Sub-humid	Maize	- Linseed/Chickpea/
		Rice (upland)	- Wheat/Sunflower
Sub-mountainous	Semi-arid	Maize	- Fallow
	Sub-humid	Maize	- Wheat

The cropping system is a variable parameter and changes with the requirement of market, farmers requirement and new crops and varieties developed etc.

For efficient use of rain water, timely seeding of efficient crops and application of balanced dose of fertilizer and manure is required. As far as possible non-legumes and legumes should be grown in equal proportion in each season for facilitating almost 100 per cent legume-non-legume rotation in subsequent season. This will help in maintaining good soil health and land productivity. Sowing of summer crops on shoulder of ridges and winter crops in furrows in paired rows with the help of ridger-seeder ensures good germination and establishment of crop stand. Under tropical conditions, success of winter crops largely depends on availability of stored moisture in the soil profile, as the winter rainfall is low and highly erratic. It has been found in alluvial soils that for stored water upto 125 mm/m, rocket plant is the best crop. For stored moisture from 125-175 mm/m mustard and for stored moisture more than 175 mm/m, gram and mustard are efficient crops. For black soils, however the corresponding crops are gram, safflower and sorghum. A litle rainfall during winter helps in reaping a good harvest of these crops. Wheat should be grown in areas with good chances of winter rains. Availability of one-two critical irrigations to dryland crops makes tremendous improvement in crop-yields.

Seed Rate Row Spacing and Fertilizer Requirement

The soils of dryland areas are not only thirsty but hungry too. They must be served with fertilizer and manure to satisfy their hunger for meeting nutrient requirement of growing crops. For obtaining high fertilizer use efficiency, half of the recommended dose of nitrogen and full of phosphorus should be drilled as basal application for summer crops. The remaining dose of nitrogen should be applied after 20-25 days only when the season is favourable. If the year happens to be a dry year, no fertilizer should be applied to the next crop. For winter crops full recommended dose of nitrogen and phosphorus should be drilled just before sowing of the crop. Introduction of a legume crop in rotation with a cereal/oilseed crop helps in saving nitrogen requirement of the later to the extent of 20-30 kg/ha. Application of FYM at the rate of 10-15 t/ha in alternate year is the most effective way not only to partly meet the nutritional requirement of crop but also to help in improving physical condition and moisture conservation of the soil. A brief account of seed rate, row-spacing and fertilizer requirement of different dryland crops is given in Table 8.4.

Table 8.4: Crops and their seed rate, row spacing and fertilizer requirement under tropical arid and semi-arid conditions

Crop	*Seed rate (kg/ha)*	*Row spacing (cm)*	*Fertilizer $N + P_2O_5$ (kg/ha)*	*Crop*	*Seed-rate (kg/ha)*	*Row spacing (cm)*	*Fertilizer $N + P_2O_5$ (kg/ha)*
Black Gram	20	45	20 + 40	Pearlmillet	5	45	40 + 20
Barley	100	25	60 + 30	Pigeonpea	10	60	20 + 40
Castor	10	90	30 + 0	Rocket Plant	5	45	20 + 0
Chickpea	20	45	0 + 50	Safflower	15	45	40 + 20
Cluster bean	20	45	0 + 30	Sesamum	3	45	30 + 0
Cotton	12	90	40 + 20	Sorghum	10	45	40 + 20
Cowpea	20	45	20 + 40	Soybean	10	45	20 + 40
Fingermillet	15	45	40 + 20	Sunflower	15	60	40 + 20
Groundnut	100	45	20 + 40	Wheat	100	25	60 + 30
Lentil	15	45	0 + 40	Tobacco	2	45	90 + 0
Maize	25	60	40 + 20				
Mungbean	20	45	20 + 40				

Horticulture in Dryland

The characteristics of the soils and climatic features of dryland areas are usually quite favourable for the production of horticultural crops namely aonla, ber, custer apple, gonda, falsa etc. There is another group of trees which can be profitably grown in the dryland area with protective irrigation e.g., citrus, datepalm, grape, guava, mango, mulberry, pomegranate, depending upon water availability and crop water needs. In good rainfall areas, these fruit plants can be raised successfully without irrigation.

The annual water requirement of some of the fruit trees are given in Table 8.5 for arid and semi-arid conditions with potential evapotranspiration of around 1500 mm per annum.

Table 8.5: Water requirement of fruit trees.

Tree	*Water requirement mm*	*Tree*	*Water requirement mm*
Aonla (*Embliaca officinalis*)	850	*Karonda* (*Carrisa carandus*)	550
Bael (*Eagle marmaloes*)	650	Mango (*Magnifera indica*)	1250
Citrus (*Citrus reticalata*)	1250	Mulberry (*Morus alba*)	850
Custard apple (*Annona squamosa*)	650	Peach (*Prunus persica*)	850
Gonda (*Cordia myxa*)	550	*Phalsa* (*Growia subinequalis*)	850
Grape (*Vitis vinitera*)	1250	Pomegranate (*Punica granatum*)	1150
Guava (*Psidium guajava*)	1150	*Ber* (*Ziziphus mauritiana*)	650

For raising orchards healthy plants of high yielding variety should be raised in pits of size 30 cm diameter and 90 cm depth dug in a grid pattern of distance equivalent to recommended spacing between the fruit plants. The samplings should be 90 cm in height. After putting the sampling in a pit, it should be filled with a mixture of soil, sand and FYM in proportion 6:3:1. Adequate amount of plant protection chemicals should also be added in each pit.

Plantation in Dryland

Selection of species of trees, shrubs and grasses is done critically based on local need, choice of people and soils and climate of the area. If soils are calcareous calcium loving plants are selected. Shrubs are preferred in shallow soils in comparison to trees. The local species are raised in large areas and exotic species having better utility are introduced on a smaller area. Nurseries of the plants are raised on farmers fields having good soil and water availability. About 20-25 thousand plants per farmer can be raised in poly-bags filled up with mixture of soil, manure and pesticide. Watering is done regularly and carefully as per the need of the plant.

Nursery plants are raised on 0.6-3.0 m high and 0.3 - 0.4 m wide bunds 4-5 m part running parallel. The space between two parallel rows can be used in raising grasses. Ameliorant and FYM be used in fairly good quantity (FYM @ 10 kg/plant) for better growth. For good establishment saplings of height 0.9-1.0 m should be selected for transplanting in pits of diameter 0.3 m and depth 0.9 m. After placing the sampling in the pit, it should be back filled with the pit soil, sand and FYM mixture in the ratio 6:3:1. Watering is done as and when required particularly during summer season. This is important in initial (1-2) years of planting. Once the plants are planted watering is done six times in 1st year at an interval of 21 days during April to May and once in the month of March and October. Subsequently it is done 3 times at the rate of 15-20 l/plant in April, June and October with good quality water. Usually good quality water is not available at the site. Tankers filled with good quality water are brought to the site for watering the plants. This is a difficult task but it is to be performed. Saline water of quality upto 150-250 me/l salts can be used for irrigating *siras, jhandi, neem* and *babool*. For protecting the plantation against grazing, about 60 cm deep trench of 70 cm top width and 50 cm bottom width is dug around the planted area on which seeding of three rows of *Prosopis juliflora* (vilayti *babool*) or *Euthoribia norutoila (thor)* is done to serve as live fencing after growth. The stems are planted in summer just before start of rainy season.

There is a wide variety of plant materials (Appendix - 5) available for plantation in watershed. The selection could be made suiting to local conditions and requirement of the community. The climatic suitability of some trees for agroforestry is reported in Table 8.6.

Table 8.6: Climatic requirement of some trees for agroforestry

Species	*Annual rainfall (mm)*	*Maximum drought tolerance duration (month)*	*Flood resistance category*
Acacia nilotica (Babool)	300-800	6-8	M
Acacia senegal (Kumbhat)	200-800	8-11	H
Acacia tortilis (Israel Babool)	100-1000	10-12	M
Albizia lebbeck (Siras)	500-2000	4-5	H
Azadirachta indica (Neem)	450-1150	Long	L
Cajanus cajan (Arhar)	400-2500	5-6	L
Cassia siames (Kassod)	500-1000	4-5	H
Eucalyptus occidentalis (Safeda)	300-760	6-8	H
Prosopis cineraria (Jhandi)	75-850	8-10	H
Prosopis juliflora (Vilayti Babool)	150-750	8-10	H

L = Low M = Medium H = High

A View of Plantation - Under Dryland Agriculture on Slopy Lands

Pasture Management

In dryland areas where cattle are raised for milk (dairying), pastures are highly economic. Most popular legume species are *Dolichos lablab, Leucaena leucocephala, Cassia siemea* and *Stylosanthes humilis*. Popular grass species are *Cenchrus ciliaries, Lasiurus sindicus, Penicum antidotal* and *Pennisetum polystachyon*. Grasses are better grown in mixture with legumes from the animal nutrition view point.

Shrubs like *Lantana camara*, Berberis which do not encourage growth of grasses are to be removed manually/mechanically from areas infested with them. Also removal of bushes of *Zizyphus, Mimase* and Butea is helpful to improve yields considerably. Contour furrowing/trenching (10-15 cm wide at 6 m interval) is useful to increase productivity of Sehima nervesham grass in semi-arid region. Also various moisture conservation practices are required for good growth of forage crops. Light harrowing is required to seed and achieve good establishment of grasses like *Dichanthium annulatum* for medium to heavy soils, *Iseilema laxum* for heavy soils in low laying areas, *Cenchrus ciliaris* and *Cenchrus setigerus* for light soils in arid region.

In Himalayan hills *Themeda urindinella*, in Nilgiris Nepier grass and in Northen hills (Shivaliks) *Chrysopogon fulvus* are found suitable. A seed rate of about 5kg/ha and spacing of 1.0 m -1.75 m for tall grasses and 0.3 to 0.5 m for short grasses is usually recommended. Since most of the Indian soils are low in nitrogen about 60 kg N/ha may be applied.

The grassed area should be fenced with locally available materials such as stones, wood or with vegetative fencing of *Thor*. Over grazing constitutes the primary degrading of grass land. Rotation grazing is, therefore, required. Also strip, continuous or zero grazing may be practiced. Cutting and stall feeding of grasses is the best but usually not liked as the majority of the people have low producing animals. Depending upon the carrying capacity of the land, number of animals should be reared. For ravines 2 goats/ha is found to be safer.

Silvipastoral Systems

These systems are suitable for degraded lands in erratic rainfall areas and involve growing of grass and legume pasture in between trees. The latter provides ground cover and prevents soil erosion. Trees suitable for silvipastoral systems should be fast growing, high palatability, good cropping ability, ability to withstand browsing and resistance to drought. Pasture species for such systems should be tolerant to drought, ability to grow under shade, easy propagation, high palatability and ability to withstand overgrazing. Some grasses and legumes suitable for different situations are given here.

Situations	Species
Marginal Soils/Desert land/Ravines	*Cenchrus ciliaris, Cliteria ternatea. Lablab purporeus. Stylosanthus humilis.*
Cultivable wastelands	*Cenchrus ciliaris, Cenchrus setigerus, Clitoria tenatea. Dichanthus annulatum, Stylosanthus humilis*
Saline and alkali soils	*Chloris gayana, Diplachue fusca, Setaria sphaccelata*
Acid Soils	*Pennisetum polystachyon, Stylo santhus guinensis*
Land affected by shifting cultivation	*Cenchrus ciliaris, Cenchrus setigerus*
Marshy lands	*Brachiaria brizantha*

The choice of trees, shrubs and grasses varies with the dryness, salinity and local requirement. Some of the commonly grown trees, shrubs and grasses for dryland and salt affected areas are given in Appendix - 6.

Water Harvesting

In dryland areas not only the total amount of rainfall is usually insufficient but its distribution also fails to synchronise with water requirement of cultivated crops, fruit trees, forests plants, shrubs, grasses etc. As a result, success of these plants depends largely upon availability of reserve supply of moisture in the soil profile and/or critical irrigation. Under such conditions water harvesting is the only way for successful agriculture. It is an age old practice called *khadin* in arid region of Rajasthan. A sketch of *khadin* is given in Figure 8.2.

Water harvesting is the technique of collecting rainfall runoff water from treated or untreated land surface and storing it in the soil profile or in a farm pond for its subsequent use for irrigation. The water harvesting is of two types: i) micro plot and ii) macro plot—also known as runoff recycling. The technique is location specific and depends upon soil, topography, climate and crops to be grown. In microplot water harvesting, runoff from watershed is stored in soil profile of the cropped area. It may be interplot or inter-row water harvesting. In areas receiving rainfall (R) nearly equal to water requirement of commonly grown crops, inter-row water harvesting is useful. In this system, ridges and furrows are made by a tractor/bullock/camel drawn ridger-seeder (Figure 8.3) and crops are planted in paired rows on the shoulder of ridges. In this technique rainfall concentrates in furrows and contains in soil profile and does not go as surface runoff. Also it helps in mitigating soil crusting problem prevalent on flat land as the soil crusting particles—fine sand, silt and clay wash out from ridges and concentrate in furrow and the crust is formed in furrows leaving the ridges free from crust. The crop sown on ridges thus remains free from crusts problem which helps it to attain good germination. Also the rainfall exceeding water storage capacity of the soil profile can be easily diverted to a farm pond for its collection and later use as

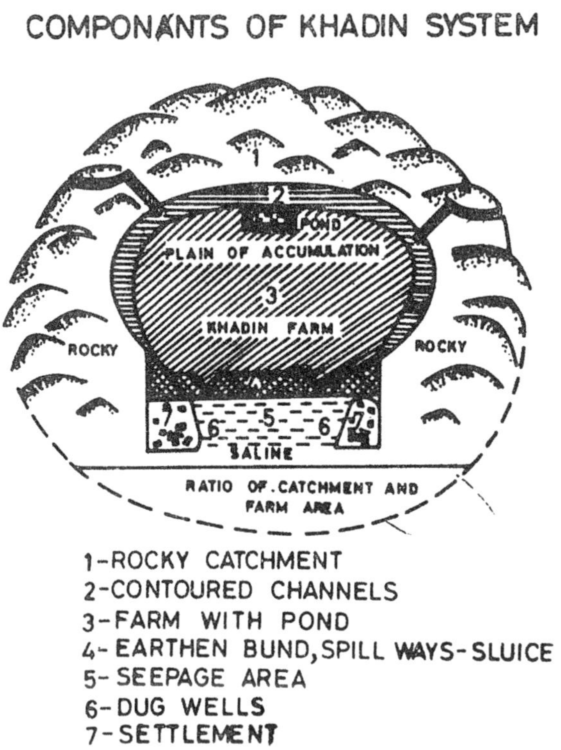

Fig. 8.2. Components of a Khadin systems

critical/life saving irrigation. For areas receiving rainfall less than water requirement of the commonly grown crops, it would not be desirable to cultivate all the available land with crops. A fruitful proposition would be to cultivate

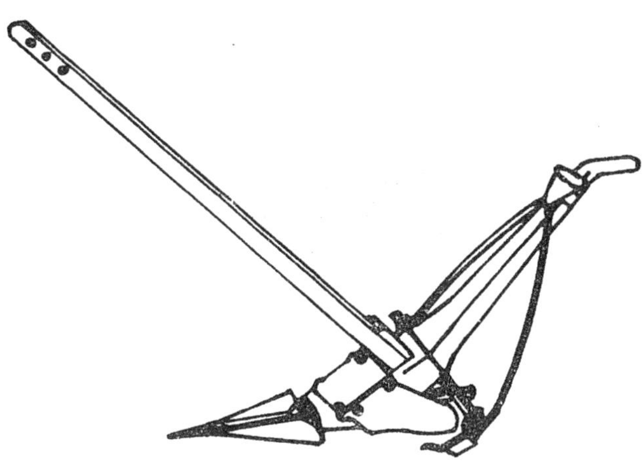

Fig. 8.3. A bullock drawn ridger-seeder

a part of the land and giving a proper slope to the remaining land such that it serves as a effective catchment for the cropped area. This technique is called interplot water harvesting technique and is more suitable for horticultural plantation.

The size of micro-catchment for tree species can be worked out by the formula

$$TS = ARC \qquad \ldots (8.1)$$

where: T = tree canopy area (m), S = runoff supplement per tree (mm) planned to be grown, A = catchment area (m^2) per tree, R = average annual rainfall (mm) and C = runoff coefficient (Table 4.1). For example a ber tree in a sandy plain having a canopy area of 36 m^2, requires about 660 mm of water in a dryland region receiving 360 mm annual rainfall. The runoff coefficient of a sandy catchment with a slope of 5 per cent is nearly 0.4. The size of micro catchment for a single plant would then be

$$36 \times 300 = (360)\,A\,(0.4) \text{ (On substitution in equation } \ldots 8.1)$$

$$\therefore \quad A = \frac{300 \times 36}{360 \times 0.4} = 75 \text{ m}$$

This means that one ber tree would need 36 + 75 ≈ 110 m land area. The value of c can be altered by making the catchment area impermeable by (i) land alteration (weeding, sloping, smoothening and compaction) (ii) chemical treatment (application of water repellent, sodium chloride solution, 4% @ 200 cc/m^2 for soil having clay more than 20 per cent) and (iii) soil cover, plastic sheet cover, etc.

Moisture Conservation

Water is an important ingredient for crop production. Therefore, it must be conserved effectively. A given amount of irrigation (I) and precipitation (P) are *partitioned* into the following components:—

$$I + P = R \pm M \pm D + ET \qquad \ldots (8.2)$$

Here the amount of irrigation and precipitation in a given period is balanced against the amounts lost in runoff from the land surface R; in changing soil profile moisture, M; in drainage below the root zone of the growing crop (D) and in evaporation from the land surface (E) as well as transpiration from the plant surface (T) ET. Therefore, for a given input of water (I&P), M can be increased by reducing R, D and ET. For decreasing R, the field should be levelled and well bunded and land should be loose at surface. Preparatory tillage for seed bed preparation is an important practice for opening up the soil and killing weeds. It has, however, been found that for loamy sand and sandy loam soils, minimum tillage (1-2 harrowing) and for sands no tilling is required. Any hard layer present in the root-zone of a crop; particularly in the surface soil must be shattered by deep ploughing with a sub-soiler/chisel plough and should be carried out once in three years. These treatments will improve infiltrability of the soil and thereby reduce runoff loss of water. The drainage below the root zone of a crop can be reduced to a minimum by applying that much amount of irrigation which is just sufficient to bring the root-zone moisture to its field capacity. Any rainfall in excess of that required for bringing the rhizosphere of a particular crop to field capacity should be harvested in a farm pond.

ET loss of water is a major component (about 75%) of water loss during an irrigation cycle. The evaporation loss of water from the land surface during a crop growing cycle could be as high as 50 per cent of total evapotranspiration. This loss must be minimized. This is possible by (i) decreasing turbulent transfer of water vapour to the atmosphere by application of vegetative mulch to land surface, creating roughness in soil surface by inter-culturing and growing trees as wind breaks and (ii) by decreasing capillary conductivity and water holding capacity of the surface soil by inter-culturing the land immediately after irrigation to make it friable and if possible by changing the texture of the surface soil to coarser side by mixing the top soil with a relative, coarser soil, borrowed from outside. Wheel hand hoe (blade type Fig. 8.4) for light soils and tined type for medium

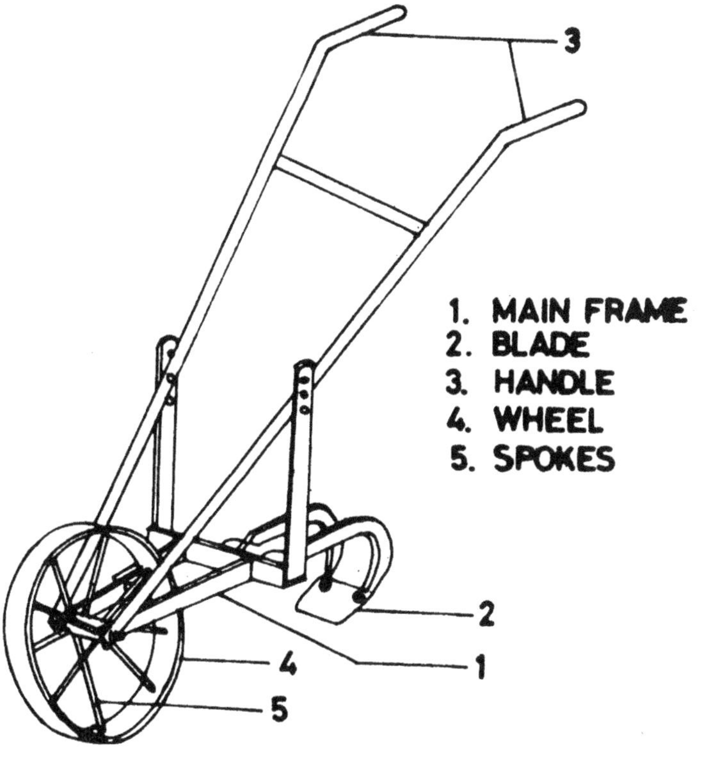

Fig. 8.4. A wheel hand hoe.

to heavy soils and blade hoe (bullock/camel drawn Fig. 8.5) or power drawn are effective implements for inter-culturing. Growing of local trees on the farm boundary helps in reducing the advective (wind) energy helping evapotranspiration. Application of straw mulch @ 2-5 t/ha immediately after withdrawal of monsoon is useful for conserving moisture and better crop stand and yield of post monsoon crops (*rabi*) raised after conserving moisture in summer fallow land or as a second crop after a summer crop. For effective conservation of rainfall in the fallow land, the land should be ploughed with a disc harrow about a month after onset of monsoon to bury the growing weeds in the ground, and subsequently after every 3-4 weeks it should be harrowed with blade/tined harrow for

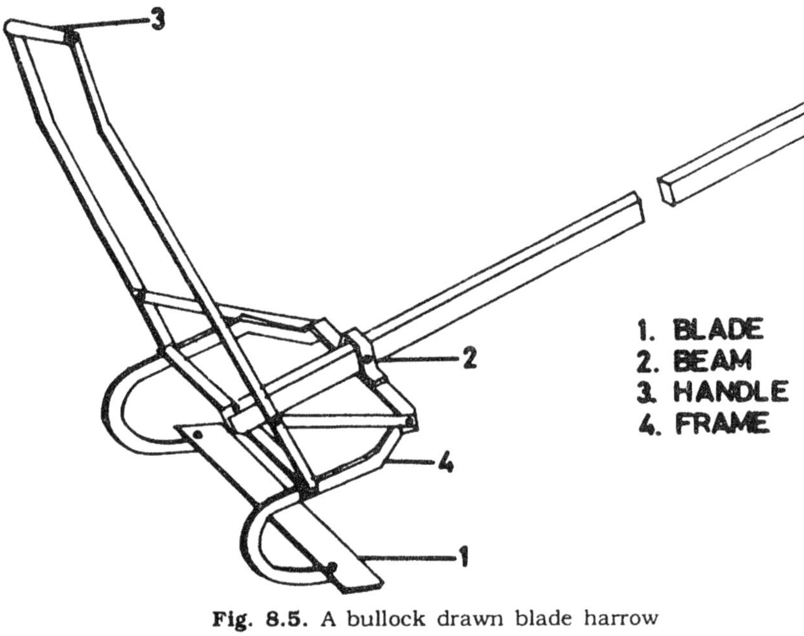

Fig. 8.5. A bullock drawn blade harrow

breaking soil crusts and creating soil mulch. After withdrawal of monsoon the land should be deep harrowed and planked for preparing seed bed for winter (*rabi*) crops. Raising of summer crops under ridge-furrow (Fig. 8.6) system in paired row on the shoulder of ridges helps to check run-off and better *in-situ* conservation of rainfall. In deep black soils opening trenches at regular interval to a depth of 0.6-0.9 m and filling with vegetative stubbles increase opportunity time for water to infiltrate in the entire profile and more moisture conservation and crop yields. Compaction of sandy soils by giving 6-12 passes with 0.5 ton roller is effective to conserve more water in the soil profile. General moisture conservation practices suitable for various soil types and rainfall classes are reported in Table 8.7. However, for a specific situation, specific practices as per the recommendation for the area should be followed. Cultivation on slopes withoutmodifyingthemproduces greater runoff, lesser infiltration. Modified as contour bunds, terraces and bench terraces, conserve rainfall effectively. Inter-culturings are required for the control of weeds and making the top soil loose, irrespective of soil type.

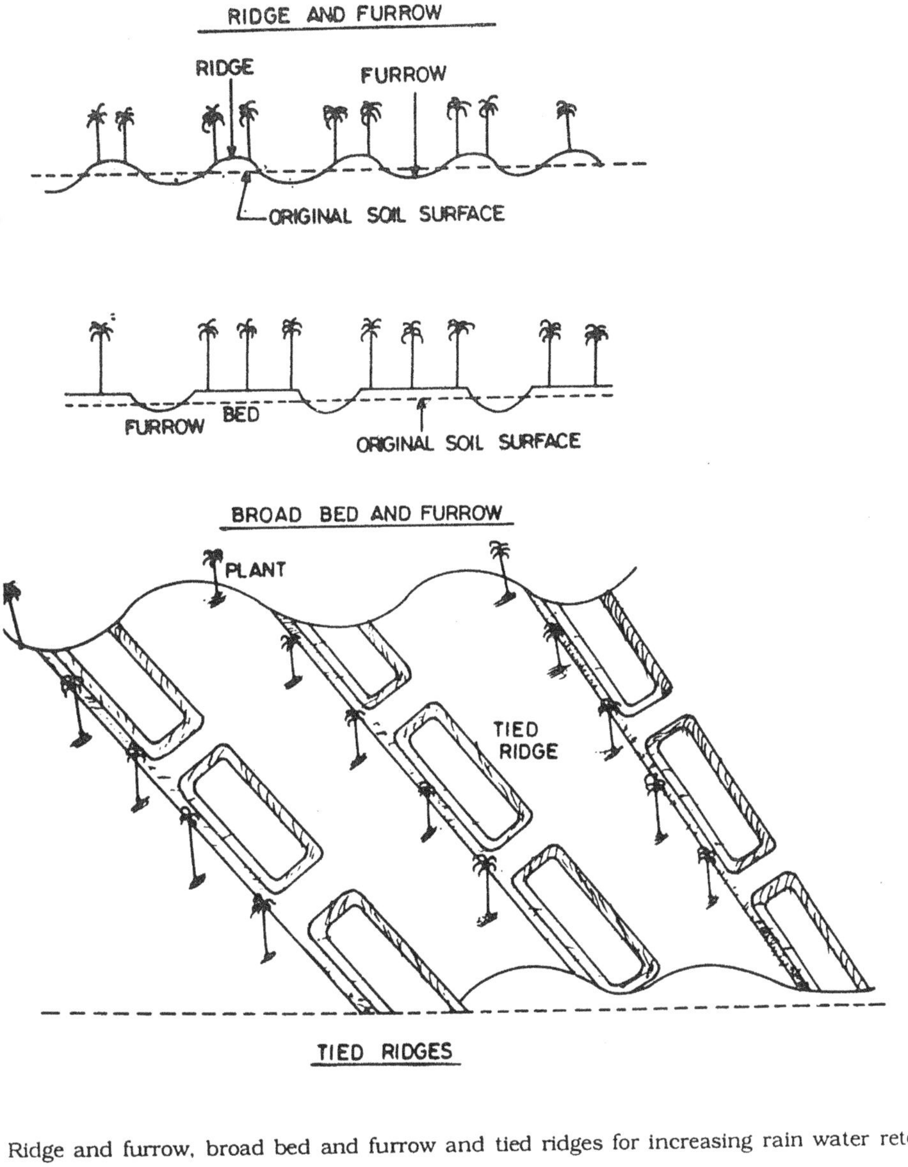

Fig. 8.6. Ridge and furrow, broad bed and furrow and tied ridges for increasing rain water retention

Table 8.7: Various *in-situ* moisture conservation practices suitable for different soil types and rainfall classes

Soil type	*Rainfall class*	*Moisture conservation practice*
Alluvial	Low	Ridge furrow planting and inter-plot water harvesting
	Medium/high	Deep ploughing, ridge furrow planting and pond water harvesting
Black	Low	Graded bunding and contour cultivation
	Medium	Graded bunding, dead furrow at 3-6 m interval, ridge-furrow planting and pond water harvesting, and tide ridging
	High	Deep ploughing, graded bunding and graded open furrow (0.2-0.3 m) at 10 m interval across the slope, bed-furrow 1.5-1.8 m width and 0.15 m depth on 0.2-0.3 per cent slope, ridge planting and pond water harvesting and tied ridging.
Desertic	Low	Sub-soil compaction, interplot water harvesting
Red	Low	Dead furrow (0.25 m deep) at 3-5 m interval and ridge planting
	Medium	Deep ploughing, ridge-furrow planting and pond-water harvesting
	High	Deep ploughing, graded border, ridge-furrow planting and pond water harvesting
Mountainous		Conservation bench terracing, flat sowing and ridging afterward and tied ridging

Tied Ridging: Tied ridging (Fig. 8.6) consists of covering the land surface with closely spaced ridges in two directions at right angles so that a series of rectangular basins are formed. The purpose of such basin is to retain rain water till it infiltrates into the soil and is suitable for level land in areas receiving high rainfall and having permeable and deep soils. Failure of a ridge particularly on slopy land can cause a series of failures of other ridges. To prevent failure of the ridges, they are constructed on grade with ties lower than the ridges so that the failure of ridges and the runoff will be along each ridge and not down the slope. It is also useful to back up the system with other measures like terraces or graded bunds. A basin lister, used for such ridging, consists of a shaft and wheels. When the implement is drawn the shovels make interrupted basins on the land surface.

Effective moisture conservation is a key to successful dryland agriculture. Every drop of rainfall should be effectively conserved *in-situ* or in a farm pond when in excess than to bring the rootzone of the soil to its field capacity moisture content. With proper conservation, harvesting and collection and use of rainfall, it is possible to improve and stabilize greatly the agricultural production from dryland areas.

Chapter — 9

Efficient Utilization of Irrigation Water

Water is the most precious and essential commodity in dryland agriculture. Every drop of water must be used most efficiently. Harvested water in farm ponds and earthen dams is very limited and usually costly. Therefore, it is of immense importance to utilize this water judiciously and as early as possible to save it from seepage and evaporation loss. The first priority should go for drinking water for human beings and animals (Appendix 6) and the remaining should be used for critical irrigation.

Critical Irrigation

Application of water for irrigation should be made on the basis of critical crop stage or when the soil has reached a critical moisture stress such that it serves as life saving irrigation to monsoon season crops or a pre-sowing irrigation to winter crops, raised after withdrawal of monsoon as a sequential crop or as a monocrop after fallowing in monsoon to serve as a critical irrigation. The improvement in crop yields due to life saving irrigation to summer crops and as a pre-sowing irrigation to winter crops is reported in Table 9.1. The results explicity indicate that the harvested water should be used for irrigating pearl millet and chickpea/mustard to achieve higher water use efficiency. This reflects that the water should be used for the most efficient water utilizer crops/cropping systems. Also the quantity of water to be applied (usually about 5-7 cm) as irrigation should be over a large area as far as possible, just enough to take care of the critical stage. When available runoff is more than the storage capacity and unexpected short drought is expected during the filling period of the reservoir the stored water should be used for protective irrigation to take care of the drought. This condition occurs very frequently in the tropical countries. Use of water in the early stage of reservoir filling may be an additional benefit of water harvesting system so long as there is sufficient water in the reservoir for later use.

Table 9.1: Effect of critical irrigation on performance of dryland crops

Location	*Soil type*	*Crop*	*Yield t ha^{-1}*	
			Without irrigation	*With critical irrigation*
Agra	Alluvial	Wheat	2.11	2.74 (5 cm)
Anand	Alluvial	Pearl millet	1.32	2.24 (7 cm)
		Tobacco	1.21	1.81 (5 cm)
Bellary	Red	Safflower	1.30	2.90
Hisar	Desertic	Pearl millet	1.32	2.24 (7 cm)
		Mungbean	0.82	1.30
		Chickpea	0.58	1.90
		Mustard	0.75	1.44
Hyderabad	Black	Chickpea	0.82	3.57 (5 cm)
Jodhpur	Desertic	Pearl millet	0.22	0.95
		Mustard	0.08	0.50
		Rocket Plant	0.62	0.67
Ludhiana	Alluvial	Sorghum	2.57	3.57 (7 cm)
		Wheat	1,92	4,11
Solapur	Red	Chickpea	0.80	1.04 (5 cm)
		Safflower	0.77	1.03

Safe Use of Poor Quality Underground Water

In some dryland areas there is a possibility of harnessing ground water which is normally poor in quality and limited in quanity. However, with scientific management such waters can be used for providing critical irrigation. Before using tubewell waters for irrigation it should be got tested for salinity and sodium hazards and other quality in Soil Testing Laboratory. Based on electrical conductivity (EC), sodium adsorption ratio (SAR) and residual sodium carbonate (RSC) of irrigation water, it can be classified into five categories (Table 9.2).

Table 9.2: Quality of brackish water

Quality of water	*Salinity (dSm^{-1})*	*SAR*	*RSC (meL^{-1})*
Good*	< 2.0	< 10	< 2.5
Marginal	2-4	< 10	Absent
Poor			
Saline	> 4	< 10	Absent
Saline sodic	> 4	> 10	Usually Absent
Sodic	< 4	> 10	> 2.5

*Good waters with RSC 2.5 meL^{-1} behave like sodic water when used in paddy-wheat or cotton-wheat cropping sequence.

The irrigation to crops should be tailored according to their tolerance level to salts/sodium, soil texture and rainfall of the area. A guideline chart for safe use of different qualities of waters for irrigation to crops is given in Table 9.3.

9.3: Guideline for selection of crops for brackish water irrigation

Soil texture	*Annual rainfall (mm)*	*Upper limits EC (dS/m) /RSC (me/L)*	*Crops*
Fine (Clay > 30%)	> 400	1.5/2.5	ST
		2.0/3.5	T
Medium fine (Clay 20-30%)	> 400	2.0/3.5	ST
		4.0/5.0	T
Moderate coarse (Clay 10-20%)	> 500	4.0/5.0	ST
		6.0/7.5	T
Coarse (Clay < 10%)	> 500	8.0/7.5	ST
		16.0/10.0	T

Crops are grouped into various categories (Table 9.4) based on their tolerance to salinity/sodicity of irrigation waters.

Table 9.4 : Grouping of crops based on their tolerance level.

	A. Salinity	
Tolerant (T)	*Semi Tolerant (ST)*	*Sensitive (S)*
Barley, wheat and date palm	Cotton, maize, mustard pearlmillet, sorghum & ber.	Groundnut, pulses, citrus, grape
	B. Sodicity	
Barmuda grass, karnal grass, para grass	Barley, wheat, oat date plam	Cotton, maize, mustard, pearlmillet, citrus, grape

Water of EC more than 16,000 µs/m and RSC more than 10 me/L can not be used for irrigation purpose.

Management Practice for Safe Use of Brackish Water

(i) ***Crop Selection***: The crop should be selected based on quality of water and tolerance limit of crops (Table 5.4).

(ii) ***Irrigation Method*** : The drip and sprinkler methods of irrigation cause less accumulation of salts and less harm to crops. In field experiment on a loamy sand soil it was found that irrigation by drip with water of EC 3 ds/m did not limit the yield of potatoes and tomatoes while that with furrow irrigation the yield reduced substantially. The sprinkler method of irrigation with highly saline water is, however, not suitable for crops like cotton, pearlmillet etc. For ber the limit of EC is 6 dS m^{-1}.

(iii) ***Application of Manure and Fertilizers***: Application of balanced dose of inorganic fertilizers and manure is useful to enhance salt tolerance limit of crop. For sodic waters nitrogen dose should be enhanced by 25 per cent and that of zinc should be doubled to mitigate the adverse effect on crops. The sodic water should be neutralized by application of gypsum @ 32 kg/acre/irrigation of 7.5 cm for each unit of RSC. As far as possible a pre-sowing irrigation to crops should be given with canal water and it is better to give one deep irrigation with canal water than mixing the same quantity of canal water with poor water and use the resultant water in number of irrigations, or giving alternate good/poor quality irrigation.

Dilution of Poor Quality Water: A crop is most sensitive to salts at its germination and seedling stage. The sodicity/salinity level of water can be lowered by diluting the water with good quality (harvested pond water) i.e. by the conjuctive use of water. The resultant salinity or sodicity of diluted water can be worked out by the following expressions:

$$EC_d = [EC_p\, Q_p + EC_g\, Q_g]/(Q_p + Q_g) \quad \text{... (9.1)}$$

and

$$SAR_d = SAR_p \,/\sqrt{1 + [Q_p/Q_g\,)} \quad \text{... (9.2)}$$

where EC, SAR and Q are respectively electrical conductivity, sodium adsorption ratio and quantity of water and suffixes d, p and g respectively stand for dilute, poor and good quality of water. In areas having paucity of good quality water, poor quality water can be used for providing critical/life saving irrigation after suitable dilution.

Measurement of Discharge

Wiers and Flumes are commonly used to measure discharge of streams and flow in channels. They are constructed across streams and channels and may be equipped with continuous water level recorders like Steven's recorders. In this method, a float rests on the water in a float well, which is connected to the main channel by a pipe or trench. With the rise and fall of water level, float in the well actuates a pen which records the water level on a clock driven paper chart.

Weirs: A weir consists of a concrete, metal or timber board with an opening of fixed dimensions cut in its top edge (Fig. 9.1). This opening is called the weir notch, its bottom edge is the weir crest and the depth of flow over the crest is called the head. Weirs may be divided into two general classes: (i) sharp crested weirs and (ii) non-sharp crested or broad crested weirs. Only sharp crested weirs are commonly used and are of 3 types (i) rectangular having a level crest and vertical sides (ii) trapezoidal having level crest and sides of the notch slopping outward from the vertical at one horizontal to four vertical and (iii) 90° triangular weir flanked by the sides of the notch slopping outward from vertical at 45° angle (Fig. 9.2). For measuring the rate of flow in a channel, a weir of known dimensions is placed across the channel and the head of flow over the weir is measured. The general formula for discharge (Q) is

$$Q = CLH^{m}$$

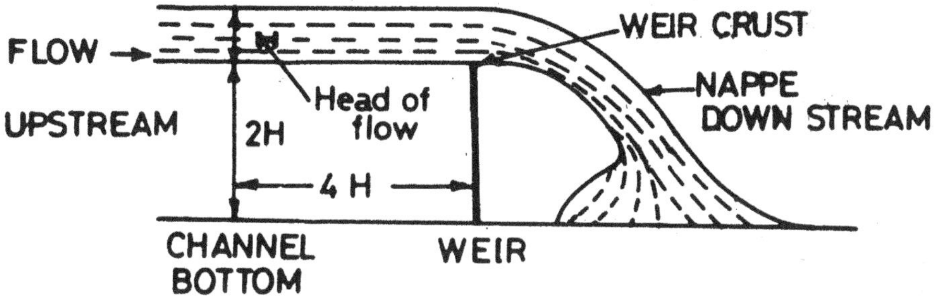

Fig. 9.1. Different terms used with weirs

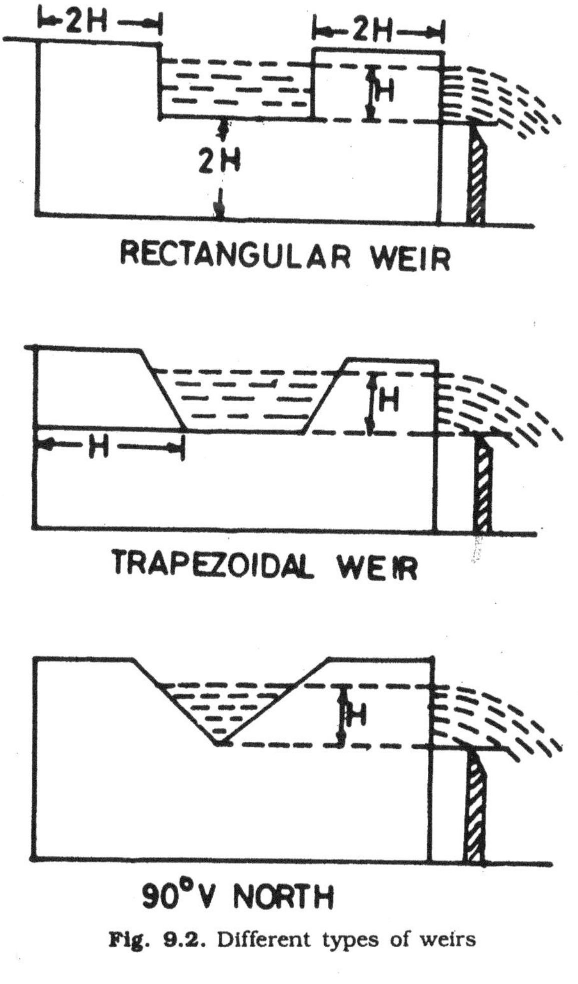

Fig. 9.2. Different types of weirs

where C is coefficient depending upon the nature of crest and approach conditions, L' is Length of the weir crest, H is head on the crest and m is exponent depending upon the shape of the weir opening. For various types of weir, the formulae for Q are

$$Q = 0.0184\ (L\text{-}0.2H)\ H^{3/2} \text{ for rectangular weir} \quad \ldots (9.3)$$

$$Q = 0.0186\ LH^{3/2} \text{ for trapezoidal weir} \quad \ldots (9.4)$$

$$Q = 0.0138\ H^{2.48} \text{ for } 90^{\circ} \text{ triangular weir} \quad \ldots (9.5)$$

where Q = discharge (1/s), L = length (cm) and H = head (cm).

The following conditions must be met to use above forumlae for weirs :

(i) The weir should be installed in a straight line at right angle to the direction of flow.
(ii) The crest and sides of the edge should be sharp edged,
(iii) The velocity of approach of water to the weir should be negligible.
(iv) The head over the weir should be measured on the upstream side at a distance of at least 4 times the head of flow over the weir. Again the distance between the ends of the notch and the sides of channel should not be less than 2 times the depth of flow of water over the weir. Similarly, the weir crest should be placed no closer than 2 H from the bottom of the channel.

Orifice

An orifice (Fig. 9.3) is an opening with closed perimeter and of a regular shape through which water flows. The orifice acts as a weir when the opening flows only partially. Orifice must have sharp edge. The stream of water coming out of the orifice is called jet. If the jet discharge into air, the orifice is said to have free discharge and if the jet is under water, it is called submerged orifice, the depth of water producing the discharge is called the head and in case of submerged orifice the head is the difference in the upstream and the downstream water levels. Submerged orifices can be conventionally used to measure small discharges. Orifices may be circular, square or rectangular. The discharge through a sharp edge orifice is given by

$$Q = C \times 10^{-3} \times a\sqrt{2gh} \quad \ldots (9.6)$$

where Q = discharge (1/s), a = area of cross section (cm^2), g = acceleration due to gravity (cm/s^2), h = head of water causing the flow (m) and C = coefficient (0.62 for sharp edge orifices).

The main advantage of submerged orifice is the suitability for measuring flows in channels having very small slopes, where it is difficult to obtain enough fall for the use of weirs. They have disadvantages in collecting floating debris and sand and silt above the orifice preventing accurate measurement.

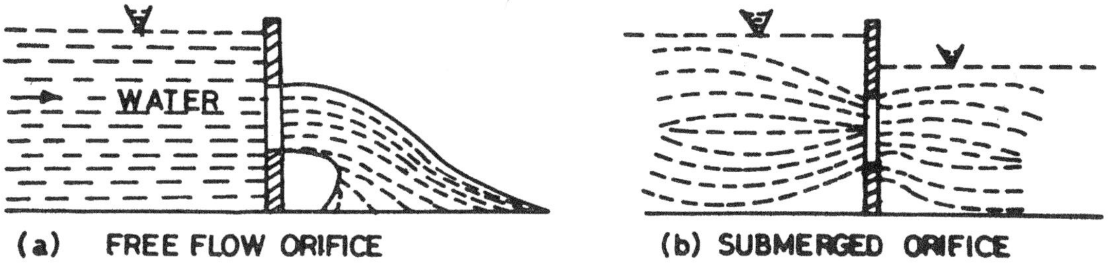

Fig. 9.3. Different types of orifices

Parshall Flume: A flume consists of three main sections (i) a converging section at its upstream (ii) a throat and (iii) a diverging section at its downstream (Fig. 9.4). The large size flumes have an approach floor and wing well at the upstream end. The floor of the concerting section is level, both longitudinally and transversely, the floor of the throat inclines downward while that of the diverging section remains upward.

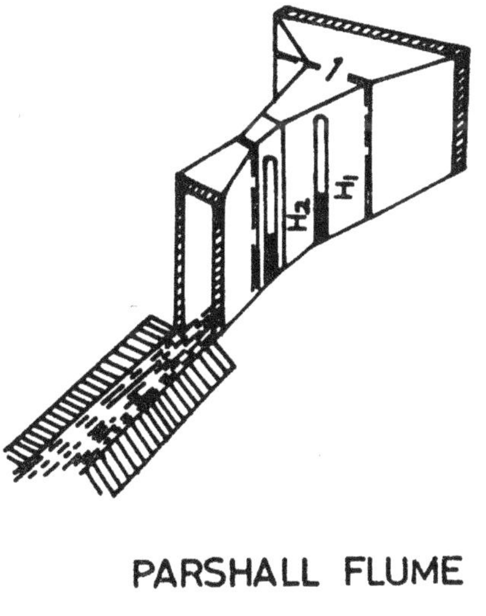

Fig. 9.4. A sketch of parshall flume

The parshall flumes are of 3 sizes to measure different discharges viz.

(i) small size flumes with throat widths 2.5–7.5 cm for measuring flows from 0.3–28.4 l/s.

(ii) intermediate size flumes with throat widths of 15.0-22.5 cm for measuring flows from 110–260 l/s and

(iii) large size flumes with throat widths of 30.0–240.0 cm for measuring flows more than 450 l/s.

Discharge through the parshall flume occurs either (a) where there is no submergence, a condition called free flow or (b) where the elevation of the water surface down stream from the flume is high enough to retard the rate of discharge, a condition called submerged flow. To determine the rate of discharge, two depth gauges (H_1 & H_2) are provided (Fig. 9.4) and from the knowledge of H_1 and H_2 and throat size of flume, Q is determined from standard tables (Appendix 7).

Methods of Irrigation

Water collected in small ponds is limited. The traditional methods of irrigation are unthinkable, as losses are higher and efficiency of utilization is low, flooding is precluded. The principle of irrigation, therefore, is to cut down conveyance and evaporation losses, and to bring the root zone to field capacity moisture. It is estimated that about 20 to 50 per cent of irrigation water is lost during conveyance from the source to the field in different soils due to seepage and percolation losses, when the channels are unlined. In areas of high (> 10 mm) evaporation rates, the closed conduit (pipes) are advisable for checking all losses.

Lining of channels will overcome or minimize seepage loss, prevent erosion in the channel, provide control against damage by rodents and convey larger quantities of water in channels of a given size. Levelled (grade < 0.2%) and smoothened fields help in uniform distribution of irrigation water. The field should be levelled gradually such that at a time the top soil should not be disturbed more than 2.5 cm in a year. However, for sand dunes and uncultivated land there is no such limit. Invariably tillering and flowering stages are critical for irrigation.

Furrow Irrigation

In this method of irrigation, the water is applied in furrows (Fig. 9.5) while the crop may be sown in furrows or on ridges. This method is usually used wherever water is in short supply and efficiency of irrigation has to be improved. But it requires a special implement to make the furrows. The land should be smooth and gentle. Furrow method of irrigation should be used to irrigate all row crops. This method has the following advantages:

(i) Water in the furrow contacts only about half of the land thereby reduces crusting of the soil and evaporation and seepage loss of water.

(ii) This method may be adopted on a wide range of natural slopes by carrying the furrows across a sloping field to check the erosion, rather than down the slopes.

(iii) This method is suitable for various stream sizes by adjusting the number of furrows irrigated at any one time suiting the available flow.

(iv) In the soils where lateral movement of water is fast, alternate furrow may be irrigated for improving water use efficiency.

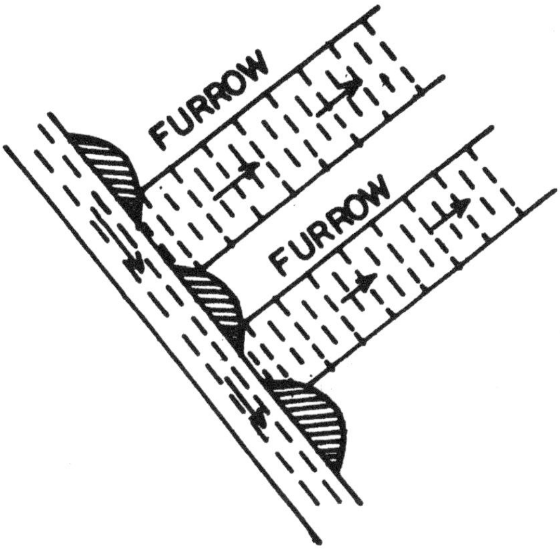

FURROW IRRIGATION

Fig. 9.5. A sketch of furrow irrigation method

Design of Furrow Irrigation System

The design of furrow involves the following steps:

(i) 4 sets, each of 3 furrows for every land slope range, not differing more than 2 per cent slope, are made. Water flow measuring flume is installed at a distance of 2-3 metres from the upstream in the central furrow of each set and another flume is installed at a distance of 100 metres from the first flume.

(ii) Water is released simultaneously in all the three furrows of a set with the help of siphons in equal quantity so that water advances at the same rate in all the three furrows. The desired quantity of water is diverted in a furrow by siphon of suitable diameter tube whose discharge is functionally related by the expression:

$$Q = 0.32\, h^{1/2}\, r^2$$

where Q = discharge (m^3/h), h = height difference (cm) between the water levels in the two limbs of the siphon and r = radius (cm) of the tube of the siphon.

(iii) When the water reaches the first flume, the rate of advance of water with time is monitored and when the water crosses the second flume, its discharge is monitored till it attains steady state. The infiltration rate of the soil is calculated by the relation.

$$(Q_0 - Q_{100}) = \text{L.B.I.} \qquad \text{... (9.7)}$$

where Q_0 = discharge at Ist flume $m^3 h^{-1}$

Q_{100} = discharge at 2nd flume $m^3 h^{-1}$

L = distance between the two flumes metre

B = width of furrow, metre; and

I = infiltration rate at steady state, $m h^{-1}$

For Example, if $Q_o = 1.5$, $Q_{100} = 0.4$, $L = 100$, $B = 0.75$, then

$$I = (1.5 - 0.4) / 100 \times 0.75 = 0.0147 \text{ mh}^{-1}$$

(iv) when the advance of water becomes noticeably slow, the stream is cut off and advance of water with time is monitored and a graph is plotted to find out the cut off length of the border, as shown in the Fig. 9.6. Let the cut off length be 140 m and the corresponding time is 43 minutes.

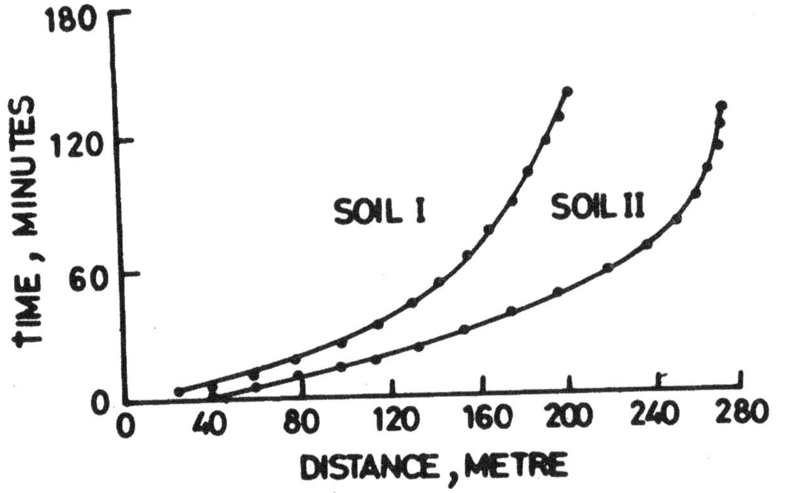

Fig. 9.6. Advance of water in a furrow irrigation test

(v) the furrow length or the discharge is adjusted such that the discharge can supply water to the soil at its basic infiltration rate after the water reaches the cut off length. For example, in this case the stream size of 1.5 $m^3 h^{-1}$ can supply water at the basic infiltration rate in a furrow length

$$Q = \frac{L \times 0.75 \times 14.7}{1000} = 1.5 \text{ or } L = 136 \text{ metres}$$

The length of furrow is reduced to 136 metres instead of the cut off length of 140 m. In case the discharge is more, it is adjusted, by adjusting the number of siphon tubes, such that they supply water in the cut off length i.e. 140 metres at the basic rate of infiltration i.e. 14.7 mmh^{-1}.

(vi) the required depth of water is supplied by application of water for the time till the water advances upto the furrow length and continuing the discharge into the furrow at the adjusted rate, if required, for the balance (remaining) time. The total time is worked out as under.

For example, if the depth of irrigation required is 75 mm for furrow of width 0.75 m and length 136 metres and the time of run (t_r) from the graph is 40 minutes, then the depth of irrigation in this time of run is

$$Qt_T = \frac{0.75 \times 136 \times i}{1000}$$

or
$$\frac{1.5 \times 40}{60} = \frac{0.75 \times 136 \times i}{1000}$$

or
$$i = 9.8 \text{ mm}$$

The balance of water to be applied is 75.0–9.8 = 65.2 mm. The time required for this depth of water application

$$Q.t = \frac{0.75 \times 136 \times 65.2}{1000}$$

or
$$1.5t = \frac{0.75 \times 136 \times 65.2}{1000}$$

or
$$t = \frac{0.75 \times 136 \times 65.2}{1000 \times 1.5} = 4.43 \text{ hours}$$

The total time of irrigation = 4.43 + 0.67 = 5.1. hours

The depth of water distribution in the furrow for the time run period (40 minute) is triangular in shape and for the remaining period it is rectangular in shape. Therefore for 40 minutes run, the moisture at the up end will be 19.6 mm and at the tail end of furrow will be zero. For the remaining time (4.43 hours), these amounts will be 65.2 mm at both the locations. Therefore, the distribution of water will be from 84.8 to 65.2 mm respectively at upstream and at the tail end.

Check Basin

This is a widely practised method of irrigation. In this method the field is divided into rectangular or square plots (Fig. 9.7) and water is introduced into each plot individually to complete the irrigation. Each plot is well levelled and surrounded by small ridges to accommodate the required amount of irrigation water within the plot. This amount in term of depth of water could be 4.5, 6.0 and 7.5 cm respectively for light, medium and heavy soils. There is usually one irrigation channel for every two rows of checks so that plots on both sides of the channel can be irrigated. The size of the plots (check basin) depends upon the infiltrability of the land and size of stream and may vary from 1.5-2.0, 10.0-12.5 and 20.0-25.0 m^2 respectively for light, medium and heavy soils for a stream size of 16 1/s. The corresponding values for a stream size of 80 1/s

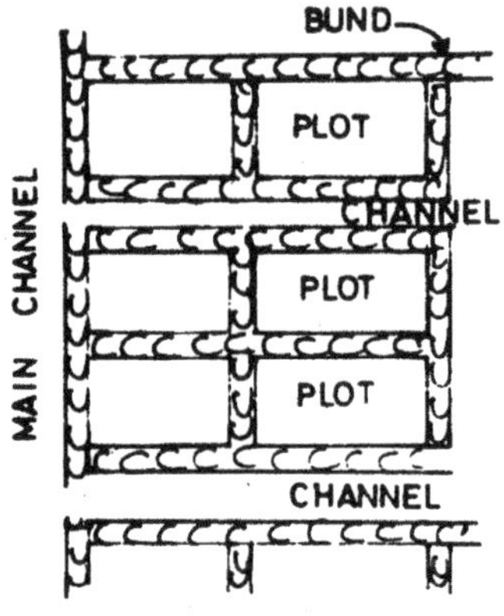

Fig. 9.7. A check basin irrigation method

are 3-4, 15-18 and 35-40 m^2. It is always desirable to experimentally determine the optimum size of plot for a given stream size and soil type. However for different types of soil and discharges of water, the following sizes of check basins and time required to irrigate the land are given below.

	Size of check basin			
Type of Soil	*Discharge 14 l/s*	*Time of irrigation (min)*	*Discharge 28 l/s*	*Time of irrigation (min)*
Heavy	200–240 m^2	20–25	300–400 m	17–20
Medium	100–125 m^2	8–10	150–180 m^2	6–7
Light	16–20 m^2	1.2–1.6	20–25 m^2	4–5

These sizes of check basins are commonly practised in the initial stage of irrigation and modified based on experience.

Problem 1: A water harvesting reservoir spreads over an area of 10 ha with an average depth of 3 metres. Calculate the area of wheat crop (130 days) that can be irrigated, if the duty of the crop is 30 ha cm.

Solution: Duty of crop = Area of the crop × water requirement of the crop.

Amount of water available = 10 × 3 × 100 ha cm

$$\therefore \text{ Area irrigated} = \frac{\text{Amount of water available}}{\text{Duty of water}}$$

$$= \frac{10 \times 3 \times 100}{30} = 100 \text{ ha}$$ **Ans.**

Problem 2: For a crop with effective rooting depth of 140 cm, calculate the irrigation interval, if field capacity moisture content is 14 per cent and permissible moisture depletion from the profile is 7 per cent (on volume basis) and evapotranspiration of the crop is 280 mm d^{-1}.

Solution: Irrigation interval $= \frac{(14-7)\times 140}{100} \times \frac{30}{28}$

$$= \frac{7\times 140}{100} \times \frac{30}{28}$$

$$= 10.5 \text{ days}$$ **Ans.**

Installation of Underground Pipeline

In areas where seepage losses are high, lands are undulating, excavation is difficult and water is to be carried down steep slopes, underground pipe conveyance system of irrigation is ideally suited. By this system of irrigation there is a saving of land (2-4 per cent), labour and water loss by evaporation and seepage. In this system of irrigation no culvert is required and there is no interference to farm operations. But this is costly system and therefore careful planning is required.

The materials used for underground pipe line system are concrete, asbestos cement, vitrified clay, PVC and steel pipes. Under ordinary conditions, concrete pipes are commonly used for irrigation water conveyance. However, if the soil contains excess of sulphate, use of concrete pipe (1 : 2 : : cement : sand gravel) and if it contains excess chloride salts, steel pipe should be avoided. Vitrified clay pipes are suitable for all conditions. Asbestos/cement pipes are more costly but are easy to install and have longer life and can be used under wide pressure range. Plastic and aluminium pipes are generally used for low discharge (small diameter) and high pressure system.

Installation of pipe lines involves selection of proper depth (> 60 cm), grade (usually existing land grade), depth and width of trench (depth 70 cm and width 60 cm), lowering and squeezing of pipes, sealing of pipes joints and back filling of the trench with care. Each pipe should be laid on firm footing, before lowering the pipes, holes for fixing riser pipes, air vents, etc. are made in the pipes at the appropriate locations. The pipes are laid with there ends placed about 1 cm apart. This gap is left for fixing a rope dipped in bitumen in the recession of pipe joints which is then compressed by squeezing the pipes. The gap between the collar and pipe is filled with a dry mixture of cement and sand (1:1). The collar is then lined with 1–1.5 cm thick plaster (1:1) and levelled off at an angle of 45^0 with the outside of the collar. The pipes are then tested for leakage at a pressure of 1 metre of water and the trench is back filled in instalments. Pipes with bell ends are laid with their bell ends facing upstream.

Structures for Underground Pipe System

Specialized structrues (Fig. 9.8) are required with underground pipeline system to control the water and protect the pipe lines, from damage . These include the structure at the inlet water control, diversion, air vents and end plugs.

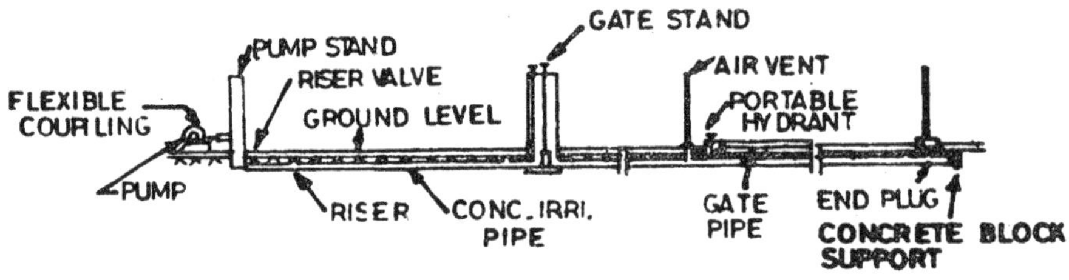

Fig. 9.8. A underground installation of irrigation pipes.

Inlet Structures: Water may enter a pipe line by gravity from an open channel or it may be pumped from a well or other sources. An inlet structure is required to develop the full flow capacity of the pipeline, to maintain water surface elevation sufficient to distribute water to the different points on the farm, to ensure that excessive pressures do not cause damage and to keep trash away from entering the pipe.

Pump Stands: A vertical pipe (≈ 60 cm diameter) extending above ground and connected to the underground pipe line system is known as stand. Its larger size permits dissipation of the high velocity stream and release of entrapped air before the water enters the pipe line. Pump stands must extend upward to a point where it will not overflow. The elevation of the water surface in the stand must be sufficient to permit the discharge through any outlet after including friction head losses in the pipe system.

Gravity Inlet: When water enters the pipe line from an open channel, a gravity inlet is provided. The water is made to pass through a screen and a sand trap to inlet structure to get rid of the problems of trash and sand in the water.

Gate Stands: These are provided to control the flow into laterals when pipe lines branch off in different directions. Gate stands also prevent high pressure and act as surge chamber. These are made in the same manner as pump stands but are provided with slide gates or valves to control flow into laterals.

Outflow Stands: These are installed in the pipelines at down steep slopes to prevent excessive pressures in the lower reaches of the pipe to serve as both a check and a drop structure. A gate may be provided between the two chambers to control the pressure of the upstream and an air vent is installed directly below the outflow stand to remove the entrapped air.

Float Valve Stands: When pipe lines are installed on steep slopes, a some-what automatic float valve is installed in the pipe to control excessive pressure down the stream. Float stands are located at an interval of about 3 m of drop in the pipe line.

Air Vents: Air vents are vertical pipe structure (10-15 cm diameter) to release air entrapped in the pipe line and to prevent vacuum. These are installed normally at all high points in the line, at sharp turns, at any abrupt change in the grade directly upstream from any structure that may entrap air at the end of the pipe line. Vents are installed at points about 200 meters apart on straight pipe line with uniform slope.

End Plugs: These are provided where any line terminates or to close the line and to absorb the pressure developed at the end of the line on account of water hammer. The plug is backed by a concrete block.

Outlets: These are required to control delivery of water to the fields at any desired locations. They should be of proper size. easy to close and open and water released should not cause soil erosion. The most common outlets consist of concrete riser to bring water from the buried pipe line to ground surface and attached alfalfa valve to control the flow of water.

Riser Valves: They are screwed valves installed on top of the riser pipes to control flow of water in the pipe line by raising or lowering the valve plate.

Hydrants: These are devices placed over riser valve outlet as means of connecting portable gated pipe to the pipe line.

Gated Pipes: These are portable pipes with uniformly spaced outlets used for releasing irrigation water to fields.

Design of Underground Pipeline

The design of underground pipeline is to determine the size of pipes and height of stand required to carry out the design discharge through the system, at an specifed pressure. The discharge velocity of a pipe line is given by the formula.

$$V = \sqrt{\frac{Hdg}{2\,fl}} \qquad \ldots (9.8)$$

where : V = velocity of flow of water through the pipeline (m/s)

d = diameter of pipe (metre)

g = acceleration due to gravity (m/s^2)

H = available head of water causing flow (m),

i.e. between pump stand and the discharging point.

l = length of pipe (m), and

f = Darcy's roughness coefficient.

The f values (friction loss) in concrete/cast iron pipe as a function of pipe diameter and various flow discharges are given in Table 9.5.

Table 9.5: f value for concrete/cast iron pipe of different diameters for various discharges.

Flow 15 l/s	*20*		f in m/100 m			
			Diameter of pipe, cm			
		25	30	35	40	45
8.5	0.07	–	–	–	–	–
34.0	0.95	0.32	0.13	0.06	–	–
79.0	4.57	1.54	0.63	0.30	0.16	0.09
91.0	5.85	1.97	0.81	0.38	0.20	0.11
102.0	7.25	2.46	1.01	0.48	0.25	0.14

The working pressure of the pipe line should not be more than 1/4th of the bursting pressure of the pipe given by the manufacturer. The riser outlets are selected to suit the flow required at diversion points depending on the method of irrigation and the crop water needs. The pressure (available) at a particular point in the pipe is equal to the permissible pressure in the pipe plus gain or loss in pressure due to down or up field slope. The height of water surface i.e. pressure of water in the pump stand is equal to total pressure loss due to friction in the pipe line plus gain or loss due to elevation difference plus loss in pressure when water enters from the pump stand to the pipe line plus free board (about 0.3 metre) to meet any unusual eventualities. The loss in head (pressure) when water enters from the pump stand to the pipe line is given by 0.5 $V^2/2g$, where V = velocity (m/s) and g = acceleration due to gravity, (m/s^2). Usually this loss is taken equal to 0.5 metre.

Example 1: The maximum available pressure in the pipe line of diameter 35 cm and length 700 metre carrying 100 l/s of discharge is 5 metre. Find out the height of pump stand. The down slope of the land is 0.1 per cent.

Solution: The frictional loss in the pipe of diameter 35 cm and carrying discharge of 100 l/s is 0.48 m/100 m. Therefore, the loss in head of water in the pipe line is equal to 0.48 × (700/100) – (0.1 × 7) = 2.66 m. Therefore, the height of pump stand = 2.66 + 0.5 + 0.3 = 3.46 m which is less than 5 metres.

Example 2: Find out the optimum size of pipe line (concrete) for carrying a discharge of 18.45 l/s. The pressure difference between the discharge point and pump stand is 2 metre and their separation is 150 m.

Solution: The discharge is given by the formula

$$Q = \text{V.A.} = \sqrt{\frac{Hdg}{2fl}}\left(\frac{\pi}{4}(d)^2\right)$$

or
$$\frac{18.45}{1000} = \frac{\sqrt{2 \times d \times 9.8}}{\sqrt{2 \times f \times 150}}\left(\frac{\pi}{4}\right)(d)^2$$

or
$$\frac{91.90}{1000} = \frac{d^{5/2}}{f^{1/2}}$$

From the table of f v/s of d and v, we can find out that for d = 0.15 metre, f = 0.009, we have the ratio $d^{5/2}/f^{1/2}$ equates to 0.0919. Therefore, pipe of diameter 15 cm is needed. For other material pipes these values should be multiplied by a coefficient given below.

Aluminium pipe	0.70
PVC pipe	0.67
Concrete pipe	0.90
Galvanized iron pipe	0.60

For local losses, the formula for equivalent pipe length is L = 40 KD, where D = diameter of the pipe in metre L = length of pipe in metre and K = a coefficient to be taken from the Table (9.6) for different conditions.

Also $$L_1 = \left[\frac{D_2}{D_1}\right]^{4.87} \cdot L_2 \qquad \dots (9.9)$$

$$\text{H.P.} = \frac{0.H}{76} \qquad \dots (9.10)$$

where Q = discharge in l/s and H = height in metre.

Table 9.6: Values of K for calculation of local losses

Local conditions	*K value*
Entrance losses	
Inward projecting pipe	0.78
Sharp cornered entrance	0.50
Slightly rounded entrance	0.23
Bell mouthed entrance	0.04
Obstruction in the line	
For long radius 90° elbows	0.25
For long radius 60° elbows	0.20
For long radius 30° elbows	0.15
For tees	
straight flow	0.1
angle flow	1.5
For gate valves open	0.00 to 0.10
For gate valves half open	2.0
For globe valves (open)	3.0 to 15.0
For angle valves (open)	5.0
For sudden contractions	0.1 to 0.5
For sudden enlargements	$1 - d_1^2/d_2^2$, $d_2 > d_1$

Selection of Pipe Diameter

Diameter of underground pipe should be selected based on expected flow. For concrete and cast iron pipe the diameter as given in Table 9.7 should be used.

Table 9.7: Diameter of pipe inlets for different discharges.

Flow (m^3/h)	*Diameter of pipe, cm* 10	15	20	25	30	40	45
2	0.01	-	-	-	-	-	-
4	0.05	-	-	-	-	-	-
8	0.18	0.03	-	-	-	-	-
18	0.78	0.11	0.03	-	-	-	-
40	3.40	0.48	0.12	0.04	-	-	-
60	7.00	1.00	0.26	0.08	0.03	-	-
80	-	1.70	0.43	0.15	0.06	0.01	-
100	-	2.60	0.68	0.23	0.09	0.02	-
150	-	5.50	1.40	0.48	0.18	0.04	-
200	-	9.50	2.40	0.90	0.30	0.08	0.04
400	-	-	8.80	2.90	1.20	0.30	0.17
800	-	-	-	10.30	4.10	1.95	0.58
1000	-	-	-	-	-	2.60	0.90

Problems in Underground Pipe Line

The following are the common problems:

(i) development of longitudinal cracks in the pipe usually at the top or both at top and bottom,
(ii) telescoping of sections,
(iii) pushing of pipes into stands, (iv) circumferential cracks, and
(v) surging or intermittent flow of water.

The first four problems are due to expansion and contraction of concrete pipes by wetting or drying conditions in the pipe lines. Circumferential cracks are caused by drop in water and soil temperature. The axial stress in the pipe line leads to longitudinal cracks. This may also cause telescoping and pushing of pipes into stand. These cracks occur normally within a few days after the pipe line is laid down. Wetting stress occurs as a result of change in moisture gradient through the pipe wall or from water conditions around the bottom and around the top of the pipe. Using moist soil for initial back fill, minimizes the longitudinal/circumferential stress. Air circulation through the pipe line should be minimized while laying the pipe line to minimize this stress.

When adequate air vents are not provided in the pipe line, air gets entrapped in the moving water and surging may occur. After a short interval, the upstream portion of the water reach becomes lighter than the downstream, causing reversal of hydraulic gradient until the water with entrapped air flows back to the pumpstand and air is dissipated. This problem can be solved by providing adequate number of air vents.

Chapter — 10

Monitoring and Evaluation

Monitoring and evaluation of a project is required to determine time scale progress, resource improvement and its social and economical impact. Monitoring denotes continuous assessment of project activities in the context of implementation schedules, the use and allocation of project inputs by targeted beneficiaries and the measurement of progress against stated goals. Evaluation refers to periodic assessment of the relevance, performance, efficiency and impact of the project in the context of its objectives. The important goals for monitoring and evaluation procedure are to;

1. establish base line information,
2. track and document physical progress and achievements,
3. develop a system for comparative analysis of project components,
4. provide a mechanism for evaluating the need for project achievements and
5. execute studies to evaluate project performance.

The major problems commonly encountered in monitoring and evaluation are

1. shortage of trained and competent staff,
2. limited and weak information base,
3. inadequate inter-departmental co-operation and co-ordination requiring information exchange,
4. overall lack of institutional commitment,
5. poor understanding of external reporting requirements,
6. non-well defined objectives of collecting, monitoring data,
7. poor standard of report preparation and
8. lack of clarity about responsibilities and obligation of different units working in the watershed.

Monitoring and Evaluation Cell

In the context of watershed development a Monitoring and Evaluation cell should be established right from the beginning at project planning stage. The task of this cell would be continuous monitoring and evaluation of various facets of watershed activities and the benefits accruing out of these developments.

Monitoring

Monitoring of the progress of development in the watershed involves regular scheduled site inspection to assess work completed and should include statement of expenditure incurred and problems and delay encountered. Monitoring progress is an essential feed back mechanism as these findings define how annual work plan should be modified or corrected. Monitoring activities use several different technologies. The most common approach is to have site inspectors recorded information on performance recording documents. These documents should be used as part of the permanent database record. In practice it is difficult to have independent field verification of the progress by actual field visits by monitoring expert team. A complementary approach is to utilise remote sensing technology, such as high resolution satellite pictures to assess physical progress. For example, site inspectors could clearly see the progress made, especially regarding forestry plantations in upland watershed areas. While these images could possibly be used quite effectively, for example to evaluate the total impact of larger projects they are probably not useful for monitoring small catchments of less than 3,000 hectares. Another approach, though not used widely as yet would be to perform field monitoring using low cost, hand held video cameras to record construction or plantation activities as they are executed.

Monitoring Results

Moditoring results also involve field inspections and supervision but rather than simply indicating physical progress, attempts to assess the effectivenss of activities in terms of survival rates, erosion control fuel-wood produced, etc.. Glosely related is the concept of monitoring benefits. Sample survey should be done quarterly or annually to assess the beneficiaries perception of the projects effect on pasture rehabilitation, improved farm and fuel-wood yields and reduction of erosion rates. Village level meeting should be held on a regular basis to elicit beneficiary response to the progress of works completed. Again monitoring results could use a combination of personal interviews, video recordings, public meetings and formal questionnaires to record information.

Monitoring results could also include more detailed long-term studies related to project progress, including information on disbursements, the type and location of erosion control structures financed and dissemination of erosion control technology through extension services. Other programmes might focus on physical monitoring for recording the effect of project investments on the rates of erosion.

Monitoring Benefits

Monitoring benefits attempts to describe the annual increments to production activities, either directly or indirectly to project investments. This monitoring function focuses on the interaction between project activities and reactions of the target populations. In watershed projects, it includes, but is not limited to, assessment of grazing, farming and other benefits. Such monitoring is usually based on periodic household surveys and should be supported for a reasonable time period following completion of the project. Typically, monitoring beneficiaries would attempt to account for changes in household incomes and relative abundance of land-based resources such as fodder, fuel, access to grazing, or water supply that could be related to project investment.

Evaluation

Evaluation of a project should be done at regular interval of time to know the impact of various development activities. The evaluation may be interim on annual basis followed by the final evaluation at the end of the project.

Interim Project Evaluation

This function involves the careful review of experience during implementation, including institutional problems, technical achievements and constraints and socio-economic benefits derived. Interim evaluation would be based on field information collected in the monitoring stage of the project. Interim reporting must go beyond an accounting of project expenditures and physical progress and should focus on the relevance, performance, efficiency and impact of the project in the context of stated goals. Ideally, both interim and overall project evaluation should be performed by an independent external agency such as a university, research institute, or qualified consultants.

Evaluation of Impact of Physical Treatments

Impact evaluation involves detailed and objective evaluation of the overall effectiveness of watershed treatments such as soil conservation, pasture rehabilitation or land management practices. Ideally the evaluation should also be carried out by a university, independent research institute, or qualified consultants. With respect to watershed projects, impact evaluation can be a very difficult task as many factors external to the project can influence the viability of investments.

Report Generation

Report generation is an important element of all monitoring and evaluation units, though typically it is overlooked. This function addresses the requirements for periodic reporting to the state on physical progress. The process is essentially manual, although word-processing facilities could be introduced to facilitate report writing. The reports would be generated directly from the data collected and systematically recorded in the watershed management information system.

Evaluation teams should also have the capacity to generate special purpose reports for external institutions such as development agencies, NGOs, research groups, or public information campaigns, This capacity should be flexible enough to address the needs of formal evaluations such as mid-term reviews, project completion studies and ex-post reviews, as well as informal management reviews. Like many other elements of the monitoring and evaluation unit, report generation must rely on the quality and timeliness of information contained within the management information system. The report and accounts generated by an evaluation team could be presented in a number of different formats such as formal written reports, diagnostic analysis or verbal and visual presentations.

Watershed monitoring and evaluation tasks are frequently cited as among the most important component of successful projects. Nevertheless, potential of monitoring and evaluation will be achieved only if it is clearly understood who is to use the results and for what purpose and if the practical limitations of available methodologies and techniques are accepted.

Economic Analysis of Watershed Development Programme

The benefits from watershed programme can be classifed as (a) protective or ecological (b) productive or economic and (c) employment. It is at times quite difficult to quantify the benefits of soil conservation programme on improvement of various facets of the environment. It is now recognised that rainfed areas, particularly arid and drier parts of semi arid tropics represent very fragile environments where ownership of individual properties rest with farmers of extremely poor resource base.

These areas have been witnessing repeated crop failure and confidence of the farmers in productive agriculture is completely shattered. From prolonged neglect and lack of investment, these areas have degraded to such an extent that these have ceased to be responsive to productive agriculture. These wastelands have to be rehabilitated and brought back in a form that these areas respond to the technological inputs to crop production. The investment in this area should be on government account. In addition, engineering and vegetative treatment should contribute to improved hydrologic regimes in term of recharge of ground, prolonged stream flow, moderation in fast flow, etc. Production or economic benefits can be quantified based on pre-project socio-economic surveys, provided the base parameter can serve as socio-economic indicators of growth studies. Expenditure on productive agriculture should be borne by the farmers. However it should be made available as low credit interest loan.

Benefit and Cost Analysis

In the simplest sense, benefit-cost analysis is no more than a budget review. It asks the question, do the benefits exceeds the cost. In the arena of public policy choices, particularly the complex issues surrounding natural resources and the environment, benefit-cost analysis is an important tool. It is useful for evaluating public interventions designed to correct perceived market and policy failures because it helps to understand the problem.

Benefit-cost analysis for a project involves a sequence of four steps :

1. Identification of inputs and the quantities needed for implementation of the project.
2. Valuation of inputs and outputs.
3. Preparation of a cost flow table using the information collected in (1) and (2), keeping in mind the project objectives.
4. Financial and economic analysis.

Input-output Identification: In economic analysis, any effect that increases goods and services desired by society is a positive effect (output) and any effect that reduces their availability is a negative effect (input). Direct inputs/outputs are those based for money in the market, indirect effects are those that are not the result of market transactions. These externalities are one reason for cost-benefit analysis. A given effect may be direct or indirect depends on the objective of the project. Fuelwood in one project is traded in the market but in another project it is produced and distributed free of charge to the poor using quota allocation system. To illustrate this point from another prospective, whether a given input is direct or indirect depends on whether or not it is paid for by the management entity of the project. If the government provides and pays for the road system required to implement a private plantation project, for example, then the road cost is not considered in a financial analysis of the private firm. However, in a socio-economic prospective, it would be considered an input and should be included in any benefit-cost analysis.

In most natural resource projects, direct inputs and outputs are the only effects explicitly considered on money terms. Most inputs and outputs relevant for financial analysis also are relevant for economic analysis. Information on direct inputs for economic analysis can be collected from available engineering and other technical studies. Categories of inputs include manpower, land, raw material, equipment, structures and civil works. Direct inputs can be classified into three classes—ecological effects, indigenous consumption and industrial uses. Depending upon the objectives or circumstances, some outputs are called ecological effects. For example a logging road improvement project is designed to reduce log-hauling cost. The project benefit is the difference between cost of log handling with and without the project. The output can be specified in term of money. In some situations, a project avoids losses and increases production over present levels. Where a hilly area is eroded, for example, and the productive top soil is being carried away it is estimated that annual production will decrease over 20 years from 0.5 to 0.0 t/ha. If a proposed project to stem erosion is designed to increase production over 10 years, the appropriate measure of project success is the increased output plus the soil loss avoided. If a higher level of production is expected to continue beyond 20 year life of the project then the benefit of production from the land beyond that period should also be considered. Under such situations, analysis over time are made both with and without a project to compare the net benefits.

It is difficult to value indirect effects in monetary terms. However, they should be identified in physical terms if possible; otherwise a detailed descriptive account of net indirect effects may be given. Several indirect positive and negative effects are reported in Table 10.1.

Table 10.1: Indirect positive and negative effects of forestry projects

Positive effects	*Negative effects*
Soil, watershed and wildlife protection and improvement of recreation and habitat, training and demonstration effects.	Pollution and negative environmental effects. Cost increase outside the project area that affects production elsewhere in the economy.
Cost savings from project activities that increase output and use excess capacity outside the project.	Infrastructure costs not included as direct costs but required for the project.

Valuation of Environmental Costs and Benefits

A difficult task is to decide which enviromental and resource effects to include and then how to quantify and monetize them—that is to assign value. No cook-book answer exists. The approach varies from project to project, but the process is more or less the same—identify the environmental effects of the project explicitly state assumptions and reach a decision. The following guidelines may be followed:

- — Effects making directly measurable changes in productivity can be valued by market prices. For example, pumping industrial effluents in to a river may cause a decline in fish production. The net change in productivity in this example can be measured in monetary terms.
- — A benefit forgone is a cost—a cost saved is a benefit. The value of an improved industrial waste water treatment, for example, can be determined to be either a direct cost or a cost avoided.
- — The analysis done using with and without project comparison.
- — All assumptions should be stated clearly—this avoids unnecessary challenges from other experts.

Surrogate-market techniques should be used to evaluate in priced environmental effects. For example, clean air may be a factor in the pricing of local housing. Analysis of price differences of housing in areas with different air quality may indicate the implicit price of the improved environment.

Benefit-Cost Ratio (BCR): For benefit cost analysis, the ratio of B/C is calculated by dividing the present worth of benefit stream by the present worth of cost stream. This ratio measures the benefit per unit of cost of investment. It is calculated as:

$$B/C = \sum_{j=1}^{n} \frac{b_j}{(1+i)^j} / \sum_{j=1}^{n} \frac{c_j}{(1+j)^j} \qquad \dots (10.1)$$

where b_j=benefit in j^{th} year, c_j=cost in j^{th} year, i = interest rate and n = number of years.

The independent project can be accepted with B/C ratio greater than 1, when discounted with the opportunity cost of capital. However, this measure may give incorrect ranking among independent projects. The ratio can be directly used to see the rise in costs without making the project attractive. It is computed by taking the present value of gross benefit less associated cost and comparing with the present worth of project economic cost. The associated cost is the value of goods and services included in the project needed to produce the output for sale or use. The project economic cost is the sum of installation cost, operation, maintenance and replacement costs and induced cost. The induced cost is the uncompensated adverse effect caused by the project.

Economic Analysis of Kandi Watershed Project

The Kandi watershed project was designed keeping in view the problem of soil and water conservation in the hill track by the Punjab Government with financial support from the World Bank. The project aimed at comprehensive and integrated development of the economically backward area through afforestation and soil conservation measures to reduce runoff and soil erosion, protection against excessive grazing by appropriate livestock development, harvesting and utilization of runoff for irrigation and development of horticulture and fisheries. The project offered a package directly to farmers and livestock owners and indirectly to all and sundry through employment generation, flood control and other infrastructural development in the area.

The data on the project costs were taken from the project-coordinator of the project located at Chandigarh. Since the objective of the project was to rehabilitate the upper catchments and to reduce the runoff and erosion from hills through soil conservation and appropriate vegetation, the actual parameters such as runoff, siltation load, ground water recharge etc; were not generated at the time of this study. This information is vital for economical analysis. However, the economic parameters relating to actual benefits were generated through socio-economic surveys in respect of improvement of forestry, animal husbandry, soil conservation, horticulture and irrigation to crops. These coefficients were used to estimate the benefits from the development works executed under the project. The returns from the benefits which has started flowing at the time of survey such as incremental fodder and grass yields from afforested/reforested areas, milk yield from improved milch animals, crop yield from irrigated/levelled lands; etc. The return from tree and fruit crops were projected using estimates based on actual survival rates. The actual cost incurred and expected returns in the future for each component were discounted to net present worth (NPW) and benefit-cost ratio (BCR) worked out at 12, 15 and 20 per cent discount rates. The internal rates of return (IRR) for each component and project as a whole was worked out and compared with the expected IRR as per the master plan. The output was evaluated at base year prices and the project cost taken as incurred at current prices.

During the period of project investments from 1980-81 to 1986-87, the index number of wholeprice of 21 agriculture commodities increased at the rate of 4.67 per cent per year. The inflation of the capital investment items would be even lower. Thus the under estimation in the evaluation parameters, which in any case are the indicators, is not very serious.

Forestry Component

In the feasibility report, it was envisaged to cover 2350 ha at a cost of Rs. 113.30 lacs. The main emphasis was on the planting of *Acacia catechu* (*khair*) and *Eulaliopsis binata* (*bhabhar* grass) in association with other species in 8560 ha land. The actual cost per hectare of the development area was Rs. 2447. The expenditure on reforestation/afforestation associated with conservation works such as vegetation and stone check dams, debris dams etc; aggregated to be Rs. 208.63 lacs.

Evaluation surveys were conducted to estimate the number of plants/ha and their final survival, the incremental grass yield (*bhabhar*) due to enclosure to grazing etc. The return were mainly from the sale of fodder and *bhabhar* grasses, *khair* trees and bamboos. The prices for these items were as per the feasibility report at the rate of Rs. 4/q for fodder and Rs. 25/q for *bhabhar*. Final harvesting of *khair* was considered in the 25th year yielding 0.14 m^3/tree with a price of Rs. 1400/m^3. The bamboo price was assumed as Rs. 3/- per bamboo. The final number of plantations were evaluated as 423/ha after 6 years of plantation and the *bhabhar* yield was 2q/ha in the 4th year and 10q/ha at full development in the 10th year. The final number of plants. tufts were estimated to be 423 and 811/ha. The incremental yield of *bhabhar*/ha from these *tufts* was found to be 0.9q/ha in the 3rd year, 1.09 q/ha in the 4th year and 6.1 q/ha at full development from 10th year onwards.

To stabilize the *choe* embankments the number of bamboo clumps planted was 1452, 804 and 1900 during 1983-84, 84-85 and 85-86, respectively. The production from bamboos clumps was assumed to start from 5th year with 2 bamboos, increased gradually to 12 bamboos at full development from 10th year onwards. The incremental yield of *bhabhar* and fodder grass (natural) was 0.1q/ha in the 3rd year, 0.2q/ha in the 4th year and projected 1.3q/ha from the 10th year onwards.

The net present worth at 12 per cent discount was computed at Rs. 252.95 lacs; with the cost-benefit ratio more than 2. It was more than 1 at 15 per cent discount but less than 1 at 20 per cent discount rate. The IRR for the watershed in the feasibility report was 14 per cent discount rate. Thus the forestry component yielded better than it was expected.

Animal Husbandry Component

The objective of the livestock component was to control open grazing in the shills and exchange of two low yielding milch cows/buffaloes or four goats for one improved cow/buffalo that would be stall fed and improvement of succeeding generation of existing animals by means of artificial insemination, thereby improving the vegetation cover and to improve the socio-economic conditions and nutritional level of the people. The expenditure on this component was Rs. 42.22 lacs where 564 improved animals were supplied, 16182 q feed/fodder were given at subsidised rates and 3562 artificial inseminations were done. The percentage of improved animals retained by the beneficiaries was 100,70,24 and 11 per cent, respectively during 1st, 2nd, 3rd and 4th year onward upto 12 years. Assuming a lactation period of 18 months, the average milk yield of these animals was 1528, 1232, 1151 and 1109 litres during the corresponding periods. The declining milk yield of the improved animals was due to the poor performance of the fodder production. To get the net incremental milk yield due to cattle exchange programme, the average milk yield was estimated to be 72.5 litres per animal/year. The population of improved female offsprings due to AI was estimated to be : 60 per cent AI success, 50 per cent ratio of females, 12 per cent death rate per young stock and the offspring in milk at the age of 4 years. The gap between the yields of the improved milch animals and the local animals was estimated to be 602 litres per year. The benefit streams from this component are given in Table 10.2-10.4.

Table 10.2: Project costs for different components (Rs. in lacs)

Particular	*1980-81*	*1981-82*	*1982-83*	*1983-84*	*1984-85*	*1985-86*	*1986-87*	*1987-88*	*Total*
	-81	-82	-83	-84	-85	-86	-87	-88	
Forestry	6.93	14.94	30.28	70.07	50.23	18.39	11.93	5.86	208.6
Animal husbandry	—	4.47	4.01	8.68	7.42	4.78	12.40	0.46	42.22
Soil conservation	—	—	12.27	11.97	4.79	6.35	23.35	3.95	62.68
Horticulture	0.08	0.21	0.09	0.41	0.17	0.64	1.23	0.27	3.10
Overheads	2.59	8.93	16.33	11.95	10.05	8.62	12.50	2.10	73.07

Source: JDC (KA). Kandi watershed and area development project, Government of Punjab, Chandigarh.

* Inclusive of overhead costs on irrigation as follows: Rs. 20.62 lacs in 1984-85, Rs. 44.44 lacs in 1985-86, Rs. 26.14 lacs in 1986-87 and Rs. 7.29 lacs in 1987-88.

The incremental financial returns from this component were computed using the net financial price of milk at Rs. 1.10 per litre which exclude the cost of Rs. 1.90/litre incurred by the beneficiary farmers. The net present value at 12 per cent discount rate stood at Rs. 3.37 lacs. The benefit-cost ratio at 12, 15 and 20 per cent rates was 1.14, 1.02 and 0.87. The IRR was 15.63 per cent which was lower than that expected (30%) in the feasibility report. This is due to exodus of improved stock and the decline in the milk yield of exchanged animals.

Table 10.3: Generalised incremental returns from various components (Rs. in lacs)

Year	Forestry	A.H.	S.C.	Horticulture	Irrigation	Total
1(80-81)	—	—	—	-0.06	—	-0.06
2	—	1.54	—	-0.17	—	1.37
3	0.05	2.09	—	-0.19	—	1.95
4	0.30	3.58	0.69	-0.41	—	4.16
5	0.96	4.55	1.53	-0.32	—	6.72
6	2.32	4.42	2.69	-0.54	—	8.69
7	3.91	3.84	3.85	-0.88	1.26	11.96
8	5.72	4.18	5.54	-0.37	6.52	21.69
9	8.31	5.13	6.44	0.30	11.15	31.33
10	10.95	6.46	7.92	1.24	15.48	42.05
11	13.66	6.46	8.16	2.28	19.97	50.53
12	15.93	6.46	8.16	2.28	19.97	50.53
13	17.93	6.46	8.16	3.52	22.92	58.99
14	18.80	6.33	8.16	4.24	22.92	60.45
15	19.19	6.23	8.16	4.47	22.92	60.97
16	19.19	6.00	8.16	5.50	22.92	61.77
17	19.19	5.32	8.16	6.27	22.92	61.86
18	19.19	4.52	8.16	6.41	22.92	61.20
19	19.19	3.56	8.16	6.41	22.92	60.24
20	19.19	2.50	8.16	6.41	22.92	59.18
21	19.19	1.33	8.16	6.41	22.92	58.01
22-24	19.19	—	8.16	6.41	22.92	56.68
25	103.89	—	8.16	6.41	22.92	141.38
26	527.29	—	8.16	6.41	22.92	564.78
27	1289.29	—	8.16	6.41	22.92	1326.78
28	3813.39	—	8.16	6.41	22.92	2850.88
29	1966.69	—	8.16	6.41	22.92	2004.18
30	577.77	—	8.16	6.41	22.92	615.26
31-35	—	—	—	6.41	—	6.41
36-50	—	—	—	1.94	—	1.94

A.N. = Animal husbandry; S.C. = Soil conservation

Soil Conservation and Engineering Component

The non-farm soil and water management development works such as land levelling, bench terracing, underground pipe-lining, etc; were carried out on 1.086 ha of land. The year wise achievements were 106, 390 and 66 ha from 3rd year to 8th year. The evaluation surveys revealed significant shifts in land use that is increase in irrigation from 11 to 49 per cent, decline in the uncultivated area from 20 to 0.17 per cent and shifts in the cropping pattern. The yield level of all important crops increased significantly. Due to the soil conservation treatment, the incremental gross income was estimated at Rs.338/ha after 1st year, at Rs. 412/ha after 2nd year and at Rs. 780/ha on 3rd year onwards. These coefficients were used to develop the benefit streams given in Tables 10.3-10.4.

Table 10.4: Net present worth (NPW), Benefit cost ratio (BCR) and Internal rate of return (IRR) for different components

	NPW (Rs. Lakhs)			*BCR at*			
Particulars	12%	15%	20%	12%	15%	20%	IRR (%)
Forestry	252.95	74.03	- 40.54	2.74	1.55	0.66	18.03 (14)
Animal husbandry	3.37	0.35	- 2.33	1.14	1.02	1.87	15.63 (30)
Soil conservation	- 3.69	—	-	0.90	—	—	10.81 (NA)
Horticulture	11.19	6.02	1.87	7.05	4.65	2.36	25.35 (NA)
Irrigation	—	—	—	—	—	—	1.22 (9.8)
Overall	-5.54	—	—	0.09	—	—	11.94 (NA)

Figures in parentheses indicate IRR expected in the feasibility reports, NA = Not available.

The project cost for this component was Rs. 62.68 lacs. The net present worth at 12 per cent discount rate worked out to be Rs. 3.69 lacs. The BCR at this discount rate was 0.90 and the IRR was 10.81 per cent.

It needs to be mentioned that the extension services to the project were grossly inadequate. With the improvement in these services to this component, the benefits are likely to increase manifold. Also due to lack of coordination with the irrigation department, the development works of the soil conservation component were not complemented with irrigation as expected; hence reduced BCR.

Horticulture Component

This component was planned to rejuvenate the existing orchards and induced fruit plantations in the project area by providing 50 per cent subsidy on purchase of plants, plant protection equipments, insecticides/pesticides and fencing material subject to an upper limit of Rs. 700 per farmer, besides subsidised interest of 2.5 per cent on loan for this purpose. This component established new orchards on 186.1 ha land with *kinnow*, mango, guava, etc;

on equal area basis. Item wise costs and returns based on input-output data of five years were projected over the economic life period of the fruit crop from the farmers point of view. The BCR at 12 per cent discount rate to the farmers for all types of orchards was more than 2 and the IRR, varied from 25 to 50 per cent for different fruit species. The farmers net benefit streams were used to work out the benefit for the project. The BCR at 12 per cent discount rate was more than 7 and IRR, was more than 25 per cent for the watershed. However, it needs to be mentioned that most of the horticultural activity was taken on the irrigated lands with no emphasis on rainfed lands.

Irrigation and Drainage Component

The irrigation and drainage component involved construction of earth fill dams to store the rainfall-runoff to prevent floods and to provide irrigation to crops and plantations. The dam and irrigation scheme was completed in 1986 at a cost of Rs. 423.15 lacs to irrigate 914 ha land (CCA) and could provide irrigation to 10 per cent (CCA) during 1st year and 42 per cent during 2nd year and was expected to cover the targeted area of 561.4 ha for cultivation due to flood proofing.

The survey estimated the gross margins at Rs. 825/ha for the additional area cultivated (without cultivation) and at Rs. 2,225/ha during the 1st year and Rs. 2780/ha from the second year onwards for the area irrigated. The incremental gross gains from the area irrigated were Rs. 1400/ha during the 1st year and Rs. 2000/ha from the 2nd year onwards. These coefficients were used to estimate the benefit stream for this component and the same is given in Table 10.3. The life span of the dam was taken as 30 years. The investment in the project started in the 5th year and the benefit started flowing from 7th year up to 30th year. The BCR yielded an IRR of 1.22 per cent which is too low to be economically justified. The excessive cost escalation and considerable delay in providing irrigation ate into its viability.

In conclusion it could be said that the Maili watershed could yield the BCR at 12 per cent discount to be 1.81 per cent without irrigation and the rate of return was more than 15.5 per cent. However, when the irrigation component was added the project was not feasible at 12 per cent discount rate.

Chapter — 11

A Case Study of Bunga Watershed

Bunga is a small village in Narayangarh Tehsil of Ambala District in Haryana (India), situated at 77° 10' longitude and 30° 42' latitutde and 300 m above mean sea level in the foothills of Shiwalik range in North India. The village had (1983-84) human population of 897 with 121 families, out of which 24 families were landless. All the farming families belonged to *Gujjar* community who were mainly shepherds by nature and were leading a nomadic existence. The landless families belonged to *Harijans*. The total cultivable land in the village was 336 ha out of which 81 per cent was being cultivated as rainfed. The average land holding was 2.8 ha with 46 per cent were owning land, less than one hectare. Only 4 families possessed more than 4 hectares of land. The village had 3340 cattle—274 goats, 820 cows, 140 bullocks and 206 buffaloes. The livestock density worked out to be 680 animals /km^2. This revealed the excessive grazing pressure of animals on village land and adjoining hills which led to denudation of hills causing heavy floods and loss of soil every year. The villagers could not construct their houses in the down plain area due to fear of floods in the event of high intensity rainfall. The whole village was situated on the one side of the hilly area adjacent to a brook.

The cycle of poverty and denudation operating in the Shiwalik foot hill region was clearly visible in and around Bunga village. The principle land use was rainfed agriculture in the valleys and degraded forests in the adjoining hill slopes. There was no facility of canal or tubewell irrigation. The small size of land holdings, highly eroded and impoverished soils, lack of irrigation and low crop yields made agriculture a losing proposition. The next possible way to survive was to have more cattle and graze them in the hills. This soon led to the disappearance of the forest cover and as a result, flash floods brought huge amount of sediment from barren hills to the plain rendering it unfit for cultivation.

Climate and Rainfall

Average rainfall of the area is about 1100 mm and 80 per cent of this is received during July and September and about 12 per cent is received from October to March. The rainfall is highly variable with coefficient of variance of 34 per cent. Frequent droughts and high intensity (105 mm/h for 30 minute duration) storms producing run-off as well as erosion are quite common and are the major concerns of environmental management. The average annual evaporation is 2300 mm. The region experiences relatively high diurnal and seasonal variations of temperature. The summer is windy, hot and dry with temperature shooting upto a maximum of 43°C during May/June while the winters are cold with temperature dropping down to 2–3°C. Frost also occurs occasionally during the month of January.

Soils

Geological formation is characterised by sand stones, grits conglomerates and clays similar to alluvial deposits of torrential streams and floods in shallow basins. The rocks are young, soft, unconsolidated and highly susceptible to erosion. The clay minerals present are illite, chlorite and kaolinite with quartz and feldspar in traces. Pink silt is dominant in the region. The coarse sediment is composed of pebbles and sand stones. The soils in agricultural watershed are generally sandy loam on the surface and loam in the lower layers. The soils in general are deep but slopes near the foothills contain pebbles and boulders. The moisture and nutrient holding capacity is low. There are patches of heavy soils varying from silty clay to silty clay loam which are calcareous and alkaline. These soils are unstable and highly prone to wind erosion. The ground water is beyond the explorable limit for shallow, dug tubewells. The dryland crop failures are common due to rainfall eroticism. The land use in general is forests on the top and hill slopes and agriculture in the valley lands. The major crops are maize, sorghum, groundnut, sugarcane, pearl millet and pigeonpea in summer and wheat and gram in winter.

Major Problems and the Programme

Land degradation and non-judicious natural resource management were the major problems of the area which was leading to misery to the people. Keeping these in mind the following major programmes were taken up:

(i) Treatment of forest catchment with appropriate soil and water conservation measures including revegetation to reduce sediment load and harvest clear water from the hilly catchment.

(ii) Harvesting the excess monsoon rain water which otherwise was going waste by constructing earth fill dam.

(iii) Conveyance of the dam water through underground pipelines system to the adjoining fields for irrigation by gravity.

(iv) Land levelling and land shaping of fields for efficient use of dam water irrigation.
(v) Demonstration and propagation of latest dryfarming technology.
(vi) Development reservoir fisheries, farm forestry, horticulture and animal husbandry.
(vii) People participation in the overall development and maintenance programme and judicious use of available resource by formation of Hill Resource Management Society. Replacement of goats with stallfed buffaloes.

Rehabilitation of the Catchment

The objective of watershed development was environmental improvement through rain water conservation and harvesting for supplemental irrigation to agricultural crops and adopt appropriate soil conservation measures with the help of local people. For harvesting rainfall runoff, a 16 metre high earth fill dam having 154.5 ha total catchment area was constructed at the cost of Rs. 21.08 lacs in 1984. The dam had capacity of 59.6 ha.m and command area of 243 ha. The hilly catchment area (46 ha) of the dam was treated by 1400 small earthen check-dams (gully plugs) and 4 big gabion structures as it was worst affected. In addition, more than 50,000 plants of *Acacia catechu (khair)*, 15400 plants of *Acacia nilotica (kikar)*, 5000 plants of *Bambusa arundinacea (bamboo)*, 1000 plants each off *Sapindus detergens* (soap nut) and *Leucaena leucocephala (subabul)* and more than 300 plants each of *Mangifera indica* (mango) and *Terminalia arjuna (arjun)* were planted in the catchment area in 1984. About 0.2 million slips of *Culaliopsis binata (bhabhar* grass) were also planted on bare hill slopes.

Details about earthen dam and its spillway

Total land area of the watershed	463 ha
Village covered	Bunga, Balwali, Bolloti, Dabkori and Kot
Catchment area of the earthen dam	127 ha (67 ha Panchayat and 60 ha forest land)
Total height of the dam	16.0 m
Storage capacity	59.6 m
Free board	2.0 m
Full lake level	14.0 m
Dead storage level	4.0 m
Live storage level	10.0 m
Submergence	10.0 ha

Rain Water Recycling

Rain water harvesting and recycling was important to ensure critical irrigation and good yield of crops and people's participation in the overall watershed development programme. Since stored water was available in limited quantity, it was envisaged to provide a protective irrigation of 7.5 cm each to wheat crop which is sown in the first fortnight of November, almost two months after the withdrawal of monsoon. The harvested rain water was conveyed to agricultural fields by gravity through four underground galvanised iron pipe lines laid 90 cm below the ground. A 90 cm diameter cast iron pipe was laid through the body of the dam with a sluice valve fixed at the lower end to regulate flow of water into underground pipe line, each again provided with a separate sluice valve. Outlets were provided at each pipe line at appropriate distance to draw water for irrigation. The command area of each pipe line and other details of the underground water conveyance system are given below in Table 11.1. Underground water conveyance system cut down water conveyance losses effectively.

Table 11.1: Details of underground water conveyance system at Bunga

Size of pipe line (mm)	*Length of pipe (m)*	*Number of outlets*	*Command area (ha)*	*Water Changes (Rs/ha)*
80	1052	5	20	3.5
103	1256	9	60	4.5
133	5431	29	168	6.0
Total	7739	43	243	-

The expenditure incurred on various equipments of the project is detailed as under:

Earthen dam (including spillway)	2.108 (Rs. million)
Water conveyance system	1.032
Catchment treatment	0.180
Choe training works	0.125
Crop demonstrations	0.048
Horticulture	0.003
Agriculture implements	0.050

Impact of Watershed Development

Crop Production

The major crops of the village in the beginning, were maize, long duration pigeonpea, groundnut, wheat and gram. Single cropping pattern was being followed under rainfed conditions. After the development of watershed, high yielding varieties of all the crops were introduced and new crops of *toria*, cotton, sugarcane, berseem and lentil were introduced under irrigated conditions and pearl millet hybrids under rainfed conditions. Among the pulses blackgram and mungbean were also given emphasis. A large number of demonstrations of various crops were laid down. In all the demonstrations, use of high yielding seeds, balanced fertilizer, pesticides alongwith latest technology of dry farming was demonstrated. For uniform distribution of limited irrigation water in the fields, farmers were encouraged to level their fields. Soil conservation measures like sowing across the slope timely hoeing and weeding were advocated. All these measures coupled with protective irrigation resulted in increase in area under crops both in summer and winter seasons and crop productivity substantially (Table 11.2). As a consequence the benefit—cost ratio at 12 per cent discount was worked out to be 1.61 for rainfed and 2.44 for irrigated crops. The results thus clearly established the financial viability of the agricultural sector. Also the watershed management programme has been able to provide stability in the yields of otherwise rainfed crops. This developed a great confidence in the farmers of the area for agricultural profession and agriculture sector.

Table 11.2: Average yield of major crops before and after the project

	Yield, t ha-1		
	Before	*After*	*Increase %*
Wheat	0.91	2.57	182
Maize	0.88	1.88	113
Pearlmillet	1.05	1.95	86
Sorghum (Fodder)	9.50	26.84	183
Sugarcane	8.30	21.29	157
Groundnut	0.67	1.23	84
Pigeonpea	0.70	1.20	71
Gram	0.79	1.04	167
Berseem	-	33.76	-
Paddy	-	3.23	-

Animal Husbandry

Livestock plays a vital role in dryland areas. Next to agriculture, animal husbandry is the most important activity for manpower employment and economic growth. According to rough estimates animal husbandry activities including distribution of animal products and fodder cultivation require about 130 man days per annum per milch animal. Livestock produce a number of vital products and services like (i) energy in the form of draft power for agriculture, rural transport and industries and dung as domestic fuel and manure (ii) food in the form of milk, milk products and meat and (iii) raw material in the form of wool, hair, hides, skins, bones, hoof and horns and number of products of pharmaceutical and industrial uses are available from the fallen or slaughtered animals.

Livestock holds much more important place in the economy of Shiwalik foothill villages and sustenance of rainfed areas. It contributes more than 50 per cent of the total income of the farmers. Density of livestock in these villages is more than three times the national average. People in these areas are pastoral nomades who keep large number of cows and goats, but are unable to feed them from their local natural resources and thus migrate along with their animals to other places in search of fodder and water. Under the watershed development programme, the grazing area was closed in parts which helped in regeneration of grasses, shrubs and trees in the closed area. To compensate the loss in grazing due to partial closure of the pasture, green fodder (berseem, sorghum) was raised in the agricultural land under irrigation to provide the needed fodder. The increased vegetation decreased soil erosion. Also, subsequent increase in available fodder both from rainfed and irrigated area, brought in dramatic changes in the animal husbandry sector (Table 11.3) in favour of buffaloes and cows at the cost of goat and bullocks.

Table 11.3: Increase in cattle wealth in Bunga

Cattle	*1983-84 Number*	*1991-92 Number*
Cows	820	1205
Buffaloes	206	561
Goats	2184	652
Bullocks	130	186
Total	3340	2604

Due to improved management, the availability of green fodder, dry fodder and concentrates increased by 144, 50 and 95 per cent, respectively, but it still fell short by 197, 116 and 167 per cent over the average actual requirement of 20 kg green fodder, 6 kg dry fodder and 1 kg concentrate per adult animal.

As a result of water resource development and associated activity, the total foodgrain production in Bunga increased from 25 ton/year to 151 ton/year, fodder from 56 ton/year to 1095 ton/year and milk production from 1231 litres to 3406 litres a day between 1984 (before) and 1991 (after).

The entire project cost was recovered in three years from increased crop production. Fishery, horticulture, farm forestry, bee-keeping, sericulture and animal husbandry were simultaneously started. After the rainy season, the grass was cut by sickles. Stall feeding became popular. Goats, the worst enemy of the forest, were gradually replaced by stall-fed buffaloes. Loaded cans of milk started going to the nearby market. There was a six to seven time increase in tree density in the forest. The mean grass yield (on which villagers heavily depend) improved from 1.9 to 8.0 quintals per hectare.

The rainwater harvested from the hills acted as a catalyst. The emergence of visible benefits from the forest watersheds made a dramatic change in the attitude of the people towards the hills. The danger of siltation of these water resources through grazing was clearly visualised by the people. After persuasion, the villagers jointly decided not to take cattle for grazing in the forest catchment lest the life giving reservoir got silted up. Villagers themselves started protecting the forest watershed. The denuders became conservators. The hills behind Bunga village started regaining the lost greenery. There was enough food for the people, fodder for the cattle and milk for the children.

The responsibility of water distribution was given to the village society. Everyone started taking interest in the programme in which the forest contractor was ousted. The lease of forest grasses was given to the village Hill Resource Management Co-operative Society (HRMS) which managed the common resources. The farmers and the foresters became friends and both started working for each other.

The society could generate a fund of about Rs. 1,00,000 each year. The revenue thus earned becomes a social fund. The money saved in Bunga was used for deepening the village pond, filling the village pond with reservoir water in summer and several other developmental activities which the village society felt were pressing. The various benefits to the society and natural resource conservation of the watershed area led to the conclusion that for the overall development of rainfed areas planning on watershed basis is the only answer.

Chapter — 12

Remote Sensing and GIS Application in Watershed Management

Proper planning is essential for the conservation and management of water and land resources of watersheds for optimum productivity. Characterization and analysis of watershed are prerequisite for this. This involves measures of hydrological, geological, hydrogeological, geomorphological, soil, land use, land cover etc. parameters and their analysis. The remote sensing technique can provide a vital tool for the measurement of these parameters. In general, the remote sensing (RS) technique be applied for:

- Delineation of watershed,
- Watershed characterization and assessing priority evaluation, problems, potentials and management needs,
- Erosion intensity mapping and identification of erosion prone areas,
- Drainage pattern mapping, and
- Evolving water conservation strategies, selection of sites for check dams, reservoirs, water harvesting structures, etc.

By using both the conventional and remotely sensed data, it is possible to evolve soil-water conservation strategies in a watershed. Such remotely sensed data alongwith geographic information system (GIS) can be effectively used for watershed planning and evaluation.

Remote Sensing

Remote sensing is defined as the science of collecting and interpreting information of an object without being in physical contact with it. It is based on the measurement of electromagnetic (e.m) energy from the sun which is reflected, scattered or emitted back by the objects on the surface of the earth. Different objects return different amounts of energy in different wave lengths of the e.m. spectrum. Detection and measurement of these spectral signatures forms the basis to identify the different surface objects on earth from air borne and space borne platforms.

Only selected portion of the e.m. spectrum can pass through the earth's atmosphere with relative little attenuation and are used for remote sensing purposes. The selected e.m. radiations employed in remote sensing include: 0.4 to 0.7 μm, (visible), 0.7 to 3.0 μm, (infrared IR), 3 to 5 μm and 8 to 14 μm (thermal infrared, TIR) and 0.1 to 30 cm (micro-wave). A typical remote sensing system consists of a sensor to collect the radiation and a platform which can be a satellite, rocket, aircraft, or a ground based stand on which the sensor can be mounted. The information received by the sensor is suitably sent back to the receiving station on earth, where the data can be reformed and supplied to the user departments either in the form of computer compatible tapes (CCT,s) or photographs/images.

The technique of remote sensing may be divided into i) active and ii) passive remote sensing.

Active Remote Sensing

A system based on the illumination of an object/scene by means of artificial radiation and subsequent collection of the reflected energy returned to the sensor. The example is microwave/radar and the system using the Lasers.

Passive Remote Sensing

In this type of remote sensing, the source of capture/detection of image or an object representing the reflection or emission of the e.m. radiation, which has a natural source. The example is general remote sensing data.

In the beginning such data were generated through aerial photograph taken from a low altitudes flying aircraft. Nowadays space satellites (Landsat TM, IRS 1A/B/C/D-PAN, SPOT, etc;) generate such data.

One of the major problems faced in the use of optical remote sensing data available from satellites are cloud cover and more soil depth to which these signals cannot penetrate. However, remote sensing data collected in microwave (MW) region of the e.m. spectrum overcomes the problem of cloud cover as MW can penetrate through cloud cover. Also MW are sensitive to soil moisture contents, water body, plantation, geological identity etc. However, its penetrating power through the soil moisture is limited (5 cm). The resolution of different satellites ranges from 56 m × 76 m (Landsat MSS) to 30 m × 30 m (Landsat TM), 10 m × 10 m (SPOT) and 5.8 m × 5.8 m (IRS - 1C/D, PAN).

Aerial Photographs and Mosaics

Aerial photographs are photographs taken by an aerial camera fitted in a aircraft. Aerial mosaics are prepared by assembling and matching of individual photographs to form a continuous photographic image of an area. The mosaics are of two types:

(i) *Uncontrolled*: in this the photographs are oriented in position by matching a corresponding image on adjacent photograph and

(ii) *Controlled*: it is compilation of scaled and rectified photograph which has been assembled to fit plotted control points. They are as good as maps. The maps have an uniform scale while mosaics may have different measurement scales. Therefore, mosaics are geometrically incorrect while maps are correct.

Study of Aerial Photographs:

The study of photo-images can be done by the following ways:

Shape: Some images can be identified by their two dimensional shape, such as railways line, road, canal, etc.

Size: From size of the image, high ways, sea, lake etc; can be identified.

Tone: The tone of a black and white photograph varies considerably i.e. from black to white. As a general rule, objects reflecting more light back to camera appear black and objects absorbing more light appear white in tone. However, the angle of the sun plays an important role in having a particular tone.

Texture: The tonal repetition creates texture of the photograph e.g. different types of forests give different textures.

Pattern: Arrangement of objects in a repeated manner e.g. a regular pattern of trees shows an orchard and not a forest.

Shadow: From nature of the shadow, objects like smoke, waterfall etc; can be identified.

Topographic position: From three dimensional view of the photographs objects like mountains, hills, sand dunes etc; can be identified.

Matting: Plant leaf disease in a particular forest or large agricultural fields gives matting appearance on photographs.

These activities mentioned above are undertaken in orderly manner for photo interpretation. A mosaic of alternate photographs is arranged on a table and covered by a glass plate and major lands and sub lands are demarked on the glass plates. Then using a stereoscope, the three dimensional view of the demarcated area is observed to get information about the various types of objects on the ground. For conformity of the various impressions of the mosaic, reconnaissance field survey is conducted and final maps are prepared.

Satellite Imageries

When a satellite passes over the earth, data are collected at the earth receiving station and are recorded on High Density Magnetic Tapes (HDMT). The satellite transmits the data in X-band (24.3) and S-band (75-100 m). These data suffer from various radiometric and geometric distortions and therefore refined by various processing techniques. False colours composite (FCC) that is imagery is created by assigning colours i.e. red to IR-Band, green to Red-band and blue to Green-band. Any colour can be generated by mixing proportional amount of these three colours.

An image is composed of continuously varying shades of colours or grey levels. Digital representation of images is governed by spatial sampling of the continuous sample area and sampling the measured intensity per sample area. Each smallest area of the sample area is termed as picture element or pixel. For a black and white image the value of the pixel is termed as grey level. The size of the pixel is termed as the spatial resolution of the sensor. A high resolution image has more number of pixels than a low resolution image and exhibits finer details of the image.

Image Processing and Data Analysis and Interpretation

Digital image processing is the digital processing of a set of inputs images using a computer. The imagery i.e. paper print of digital data is assessed visually based on colours assigned to different objects such as red to vegetation, blue to water, grey/white for sand/fallow land etc. On the other hand the digital data recorded on e.m. tapes are analysed on sophisticated computer using image processing system and relevant soft-ware packages.

The interpretation is the identification of different objects in the scene like stadium, park, canal etc. Canal, roads, and rivers are linear segments while parks, water ponds and stadium are the compact segments. Other features like the ratio of the length to breadth, the average breadth of the segment, the class of the segment, the dominating class of the adjacent pixel along the segment are selected to distinguish between canals, roads, and other linear segments. Similarly features like the percentage of water pixels in the segment, the ratio of grass pixels to other class of pixel in the segment are used to distinguish between parks, stadium and other compact segments.

Geographical Information System (GIS)

The space sattelites have been providing a flood of data about the earth's surface everyday. These data alongwith data generated by earth bound surveys and cartographic information are required to be linked and analysed for various planning purposes, which is not possible manually. However, it is possible by GIS. Remote sensing linked with GIS is an important tool to generate digital database which stores enormous volume of information and makes updatation and analysis easier.

GIS may be defined as a computer aided support system which involves capturing, storing, manipulating, analysis and displaying of both spatial and non-spatial data. It is specialized data system that preserves locational identity of the information it records. A digital computer provides the basis for storage, manipulation and display of large amount of data that have been encoded in digital form. In essence, a GIS consists of a series of overlays of information for a specific geographic region. These overlays may depict raw data or may show thematic information (such as soil, land use, topography, geology, drainage) but they must store common geographic qualities (including the common geographic coordinate system) that permit them to be merged into a single system that allows use of the varied data as an ingredient unit. Although these data can be visualized as a set of superimposed images (Fig. 12.1), they are infact stored in digital form suitable for retrieval and analysis. Specialized computer programmes and display equipments permit rapid analysis of data with a flexibility not possible by manual methods.

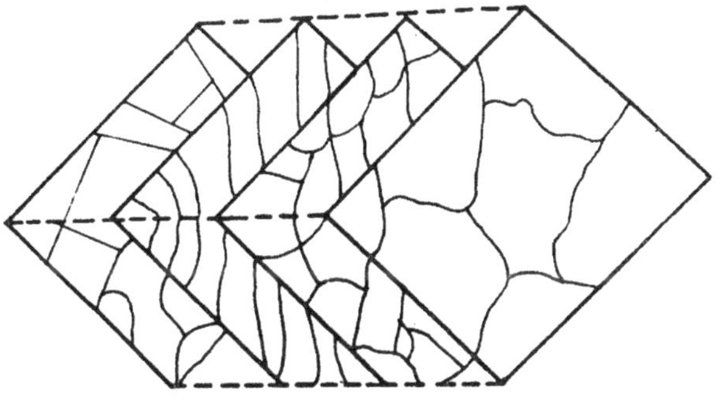

Fig. 12.1. A GIS portrayed as a series of overlays

A GIS has different kinds of information. A detailed planimetric base accurately records position of primary features of the land scape. This base forms the locational frame work that permits accurate positioning and registration of the various overlays of thematic data such as forest types, rivers, highways etc; that form the GIS. It may also include cadestre—a record of land parcels, and depicts each unit of law defined by its legal description ownership. A third element consists of the varied thematic overlays that describe physical features, streams, rivers, soils, geology, etc.

GIS has Three Important Components

Computer hardware: It consists of the main computer system or central processing unit (CPU), the terminal key board and visual display unit (VDU), digitizer, disk drive, tape drive, potter, printer, etc. The CPU is linked to a disk drive storage unit which provides the space for storing data and programmes. The digitizer is used to convert data from maps and documents into digital form and send to the computer. A plotter (output device) is used to present the results of the processing and a tape drive is used for storing data. The user interacts with the computer and the peripherals (a general term for digitizers, printers, plotters and other apparatus linked into computer) via. terminals.

GIS software modules: The software package for GIS consists of five basic technical modules, namely

— Data input, editing and verification,
— Data storage and database management,
— Data analysis, modelling and cartographic manipulation,
— Data output and presentation,
— User interface/interaction.

Some of these packages are Arc/INFO, IDRISHI, GRAMS, etc.

A proper organization context: GIS is required to include and organize the information about position (possible topographical connections) and attributes of the objects recorded. All topographic data can be reduced to three basic topological concepts—the point (a well), the line (a highway) and the area (a forest). A GIS-map is a set of points, lines and areas (polygone) that are defined both by their location in space with reference to a coordinate system and by their non-spatial attributes, Spatial data are concerned with the geography of earth surface features e.g. the map of a town showing the streets, buildings etc. while the street name address and population details are the attribute data (non-spatial data).

Geographical data are represented in two different ways. Most GIS represent the map in Vector or Raster format.

Vector Form

In this form of representation the features are defined by a series of points that when joined with straight lines, form the graphic representation of the features. The points are given x, y coordinates based on the cartographic systems being used in GIS either latitude or longitude. The attributes of the features is stored in a traditional database management software programme.

Raster Form

In this form of representation the graphic representation of the feature, and the attribute details of the features are merged and put in one cell. Typically the given geographic feature is superimposed with a mesh of grid cells (Fig. 12.2) and the definition of the graphic feature and the attributes are given to each cell. Thus a soil map represented in raster form will have an identifier number for the cell indicating the geographic details of the feature and the soil type and soil depth etc; is given to this identifier which forms the attributes of the feature. The use of number makes it possible to transform geographical characteristics using mathematical functioning.

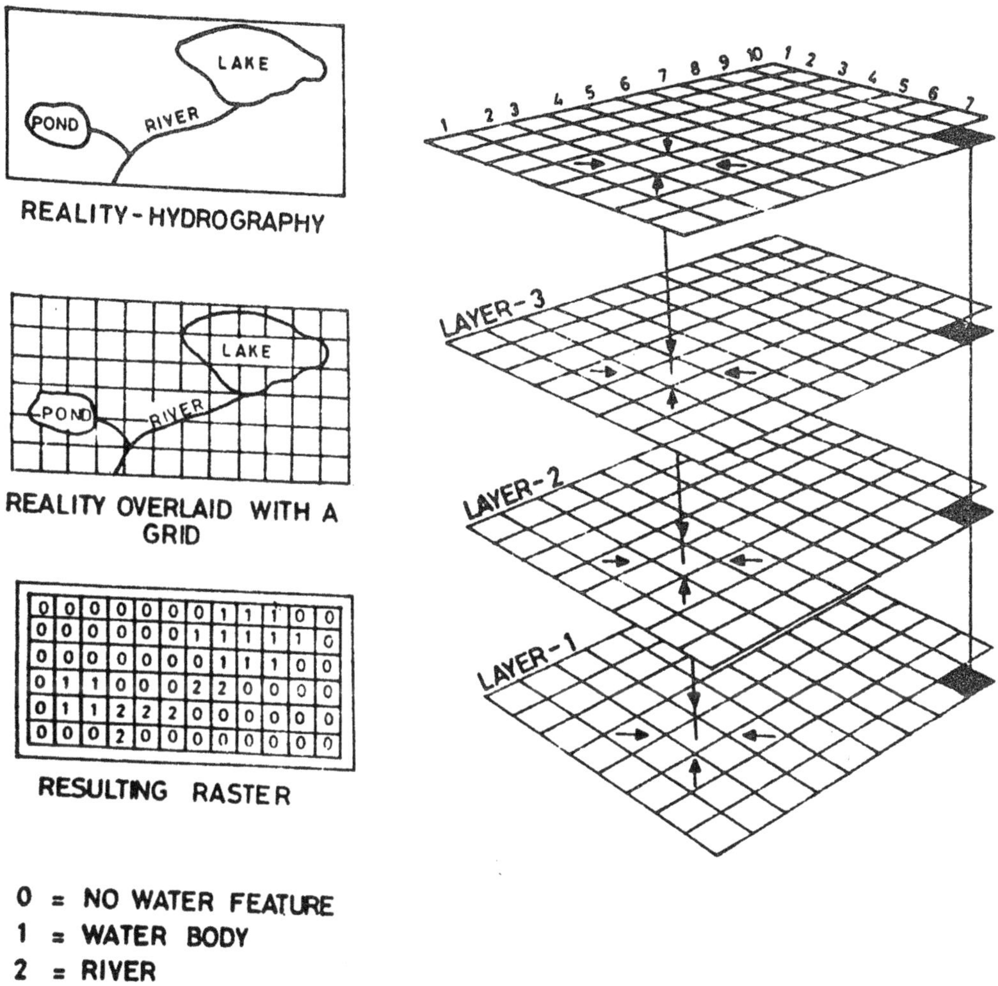

Fig. 12.2. Raster representation of data in GIS

Thematic Map Composition

The thematic map should be prepared for each of the theme viz; soil-topography, water ways, crop growth area, slope, social and economical status, etc. There are four methods to compile thematic maps;

(i) Direct transfer of the data which are on the correct scale to Mylar overlay of the base map. The data which are on a scale other than the mapping scale are rescaled to the mapping scale manually or photographically.

(ii) Themes with simple legend are rescaled using an optical Pantograph which is a kind of reflecting projector.

(iii) When the base maps are complex, manual re-scaling is too time consuming and full of certain omissions. In such cases, the scale is changed photographically.

(iv) When the base map is cadestral, the data map should be prepared on the base map scale based on field survey.

All classifications developed for the GIS needs to be expressed as numeric codes to facilitate computerized handling of the data. Classification represents qualitative, quantitative or descriptive data. In some cases classification should be developed which operates in different levels (Table 12.1).

Table 12.1: Classification of some socio-economic aspects

Parameter	*Total Marks*	*Distribution of*	
		Population	*Marks*
Concentration of SC population	25	< 25	25
		25-50	18
		50-75	12
		> 75	6
Concentration of ST population	25	< 25	25
		25-50	18
		50-75	12
		> 75	6
Literacy	25	< 25	25
		25-50	18
		50-75	12
		> 75	6

The available data are acquired in suitable digital formats and entered to the computer using the key board. The key board entry may be replaced by file transfers. These files are available from various data generating sources. Automated scaling of the data using digitizer replace manual digitizing. Nowadays digital image processing soft wares are available. The general data base required for a reconnaissance hydrological study of the water resources is shown in Fig. 12.3. The hydrological time series is limited to rainfall and evaporation. Normally the discharge is recorded at the various rainfall gauging stations.

The retrieval function is illustrated in Fig. 12.4 where the reports on the location of the rainfall stations is depicted. The various structures of the data of the general database obtained from RS exists in raster formats i.e. as photograph or imageries. Mapping units in the form of closed polygons may be either in raster or in vector format (Fig. 12.5). Each mapping needs a description and attribute tables. The description of the rainfall stations with respect to point data (total rainfall), there coordinates (location) and attribute values are given in Table 12.2. A rainfall station is retrieved by entering the name of the station and the mean rainfall and the location (measured along x and y direction) in term of x, y coordinates (Table 12.3). The management aspect of the data base within GIS is described in Fig. 12.6.

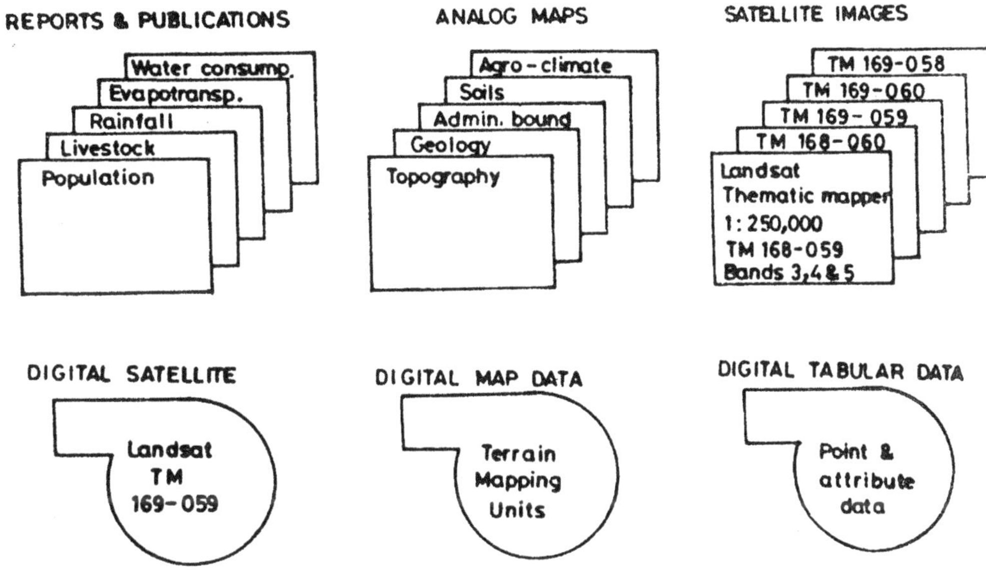

Fig. 12.3. Principle of a general database

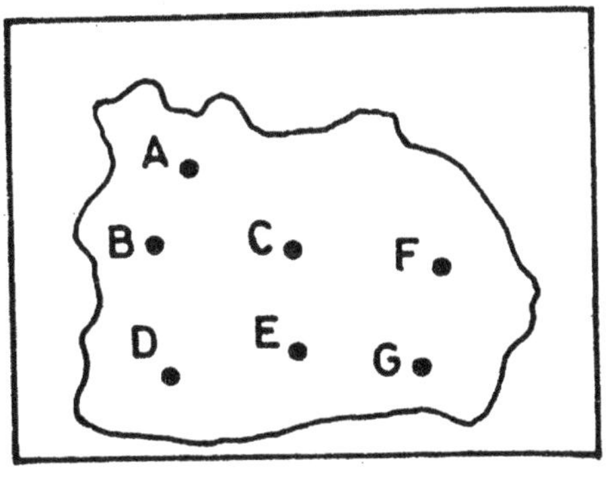

Fig. 12.4. Location of rainfall stations

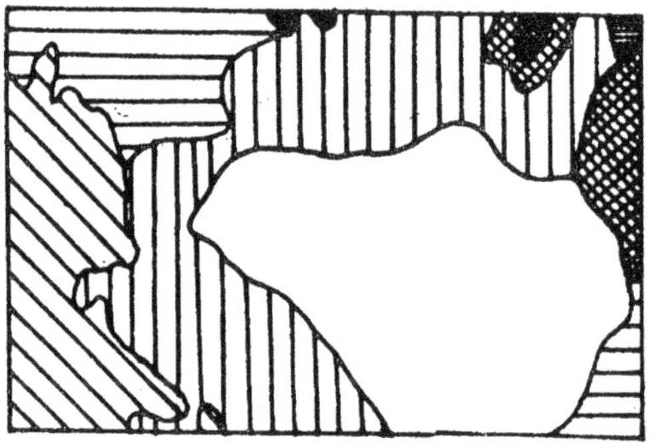

Fig. 12.5. Map data in vector/raster form

Table 12.2: Description of rainfall station

Location	*Code*	*Description*	*Lithology*
A	BH-1	Fluvial	Alluvium
B	BH-2	-do-	Colluvium
C	IG	Intensives	Acidic granules
D	IU	-do-	Ultramatic
E	B	Basement	Undifferentiated
F	V-1	Volcanic	Basalt
G	V-2	-do-	Trachyte

Table 12.3: Retrieving of rainfall stations

Station	*X*	*Y*	*Rain*	*Elevation*
A	210	520	725	212
B	95	445	360	327
C	525	495	426	401
D	100	75	511	525
E	460	130	302	393
F	840	390	417	283
G	800	60	600	437

There are several methods or algorithms to manipulate data in the spatial domain. The operations usually take place during input data stage—in a single data plane or combination of two or more data planes. The original spatial data set may have to be transformed into a map showing distribution of a hydrological variable to be used for further analysis or as input in a distributed model. Interpretation of point data is perhaps the most common operation on single data plane. Since rainfalls and ground water levels are measured in rain

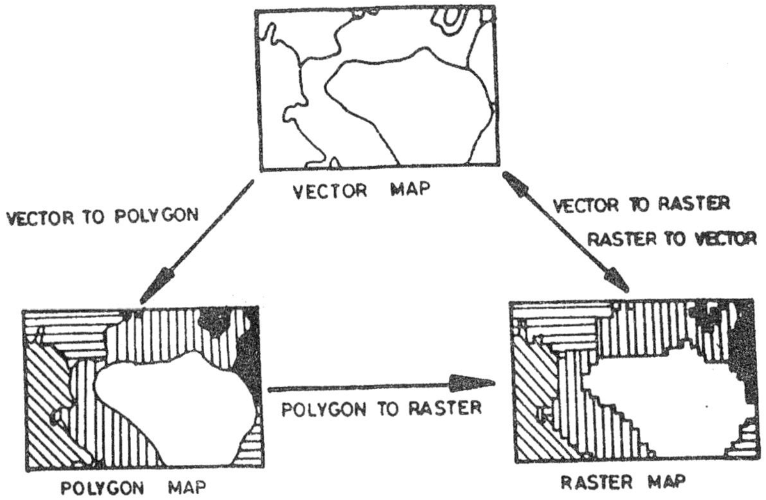

Fig. 12.6. Data management aspect within data base

gauges and in wells, point interpolation scenes must be a part of the available software. A simple but effective GIS operation is the creation of distances around points or lines such as a stream network (Fig. 12.7). Such distance zonation can be quite useful in studying yields of wells in hard rock areas as a function of distance from lineaments and for network analysis.

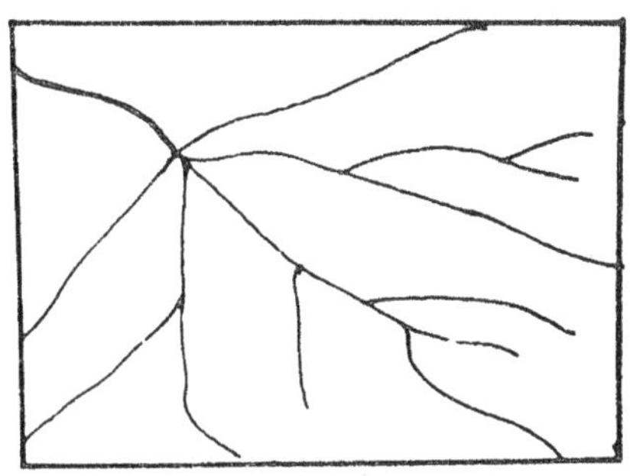

Fig. 12.7. Stream network

Interpolation and filtering techniques have many applications to transfer data. These include the sequence: interpolation from digitizer contours and points to obtain a digital elevation model (DEM), followed by filtering to arrive at a slope map or a flow path map. The latter requires neighbourhood operations or a special computer programme and is based on the down slope direction only as governed by the topography. A more complex operation on a single data plane is the creation of an insolation map whereby the direct radiation received at various time intervals is calculated from a DEM in mountainous terrain with varying lengths of shadows. Cover classification from remotely sensed data is an example whereby a single data plane is created from several data planes with spectral data.

RS, GIS and Watershed Planning

Because of the synoptic view and repetitive coverage, remote sensing has become a potential tool for effective planning of watershed development. The various watershed attributes data required for planning and the capability of remote sensing to acquire them are reported in Table 12.4. Using remote sensing data various theme maps: land-use/land cover, soil distribution, geomorphology etc; can be prepared.

Table 12.4: Watershed attributes data for planning and capability of remote sensing

Sr. No.	*Attribute*	*Attribute Parameter*	*Relevance*	*RSC & Sensor*
1	Size	Area	Average Rainfall and R&S	IRS, Landsat-TM
2	Shape	Geometric form Shape index Form factor Compactness	R&S	-do-
3	Physiography & Slope	Mean elevation Average slope Geomorphic units Total relief Relief length ratio	-do- Surface storage	AP SPOT
4	Drainage	Drainage pattern & density, Stream order & main stream length	R&S	SPOT
5	Geology	Rock types, Stratification	R&S Ground water Structure foundation	IRS, Landsat-TM
6	Soil	Soil class, suitability, capability,	R&S Land degraded & Wastelands production potential	AP, SPOT IRS L-2
7	Land use & Cover	Present land-use, Waste-land, Surface water	R&S Land reclamation	IRS Landsat-TM
8	Ground water	Potential, Quality	Recharge, Irrigation	-do- SPOT
9	Climate	Precipitation, Temperature, Wind velocity, Net radiation Evapotranspiration	R&S Production	—
10	Socio-economic condition	Human & animal population, Land holding, Present management	Watershed sustainability for food, fodder, & fuel, Social acceptability programmes, etc.	—

* RSC = remote sensing capability, R&S = runoff and sedimentation.

Resource Data Base Generation

For successful planning of watershed development programme, availability of upto date and reliable information both of spatial and non-spatial is essential. Primary resource maps required for any watershed programme are:

— Land use/cover
— Soil type/association
— Vegetation/forest
— Geomorphology
— Geology
— Topographic (contour)
— Stream network
— Surface water.

Primary (basic) thematic maps are integrated on GIS platform to device secondary (utility) maps to help in planning decision. Secondary maps are devised by combination of two or more thematic maps of chosen parameters. Some of the utility maps are:

— Soil erosion
— Land degradation
— Land capability
— Wastelands
— Surface runoff characteristics
— Ground water potential.

Watershed Characteristics

Watershed characteristics such as soil type, physiography, hydrogeology, geomorphology, etc. determine its response to hydrological process. For example the rainfall-runoff depends on soil slope, soil porosity, antecedent soil moisture and vegetation. The unit hydrograph (drainage-time response curve for a unit time duration) is one of the most important characteristics curve that reflects on the rainfall runoff response of the watershed for a given rain storm. The hydrograph is affected by characteristics of the drainage basin such as its shape, size or drainage density, land slope, land use, soil type and its infiltrability, etc. besides the amount and intensity of rainfall. Drainage density (length of the stream channels per unit area) of the watershed basin is of vital hydrological significance. A more sophisticated measure of drainage pattern is the bifurcation ratio, a measure of the amount of branching of a stream network. It is obtained by dividing the number of streams of one order by the next higher order. Similar to the mean bifurcation ratio is the stream length ratio derived by dividing the total stream length of the one order by the next higher order. Similarly, the basin morphology is the shape and relief. The shape of the drainage basin is determined by Hortons form factor which is ratio between drainage basin area and the longest axis of the drainage basin. The inferences that could be from different hydrogeomorphological factors are indicated in Table 12.5.

Table 12.5: Geomorphological factors and derived inferences

Factor		*Inference*
Shape factor	–	Hydrological behaviour of basin
Drainage network	–	Lithology and structure of the basin
Stream order No.	–	Size of the contributing watershed
Bifurcation ratio	–	Geological control to produce runoff
Stream length	–	Unit hydrograph lag
Basin area	–	Relation of stream length & discharge
Drainage density	–	Permeability of the area & sedimentation
Length of overland flow	–	Average horizontal distance between all the watersheds & streams within the basin

Remote sensing provides an important tool in determining all the factors listed in the table 12.5.

The practical watershed management problems are of two types:

(i) to estimate runoff and sediment yield for the design of a check dam for soil conservation and water storage in the watershed.

(ii) to integrate and judicially plan watershed resources for maximum benefit at minimum environmental, ecological and socio-economical costs.

Runoff and Sediment Yield Estimation

Remotely sensed data are used to study various hydrological parameters and thereby water balance, The precipitation can be measured using data of ground water radar, visible and infrared satellite images and microwave satellite data. Interception, another component of the water balance, depends mainly on vegetation characteristics can be known from vegetation ground cover. Since infiltration is more dependent on soil characteristics, the use of remotely sensed data is limited. It should be determined experimentally. Similarly, the soil moisture should also be determined experimentally due to limitation of remote sensing. The evapotranspiration can be estimated using canopy temperature and albedo images of the ground obtained from satellite.

The estimation of total and peak runoff is possible using the curve number method and Rational formula (chapter 4). The input data required are derived based on various geomorphic description of the basin and soil information which can be assessed from satellite data. The satellite data can also be interpreted to device a number of thematic maps on land use, soil, vegetation, surface water, snow cover, stream network, land form etc; which when combine with conventionally measured climatic and topographic parameters provide inputs to various runoff models.

One of the important parameters in urban watershed models is the percent of impervious area (IA) that has strong influence on runoff coefficient (C). This can be calculated directly from the Landsat data using the expression:

$$\% \text{ IA} = \exp \left[1.45 - 2.16 \frac{(\text{band } (0.8 - 1.1 \ \mu\text{m})}{(\text{band } (0.5 - 0.8 \ \mu\text{m})}\right. \qquad 12.1$$

For knowing the IA of different categories of land cover, the average values are weighted by the percent of the area assigned to the particular land cover class and summed to obtain the weighted impervious value. After classification of IA, ground survey can be carried out to establish how they react to rainfall events especially with regard to routing of storm water.

For the sake of illustration of the simple type of modelling in GIS, the Rational formula (Eq. 4.7) is taken for modelling. For determining C and A in GIS, a digital map with the series of many small catchment areas is required. The catchment areas and stream lengths are listed as attribute values in the data base (Fig. 12.8), as data capture; vector map and tabular data. A map with cover classifications is derived by remotely sensed data, which is shown in figure 12.8 as capture data: raster map and tabular data.

In order to determine the cover class per catchment, required for the determination of C, both maps are overlaid. Since the cover map is in raster format, the simple way is to first convert the vector map in raster format. Each cover class is reclassified as a look up table (LUT). The area weighted average runoff coefficient is calculated in the data base, using table operation. The result is a list of C and A values per catchment. The time of concentration and the rainfall-intensity duration could be estimated as described in chapter IV. After that the designed product is found for each catchment by table operations. The results are printed out and if required a map can be prepared with the locations of the catchments and a label with designed peak flows. In GIS all these operations can be done within the internal data base. However, the frequency analysis is usually done outside the GIS and the function is imported. The basic data needs are:

(i) Digital remote sensing data to prepare a cover map,
(ii) A table for converting cover types into peak runoff coefficient.
(iii) A topographic map,
(iv) Short duration rainfall data for a long period.

The data requirement (ii) is perhaps the most difficult. The conversion of thematic field data to hydrological variables is not difficult. Often the relationships are too complicated to be treated in a simple fashion for inclusion in the hydrological model.

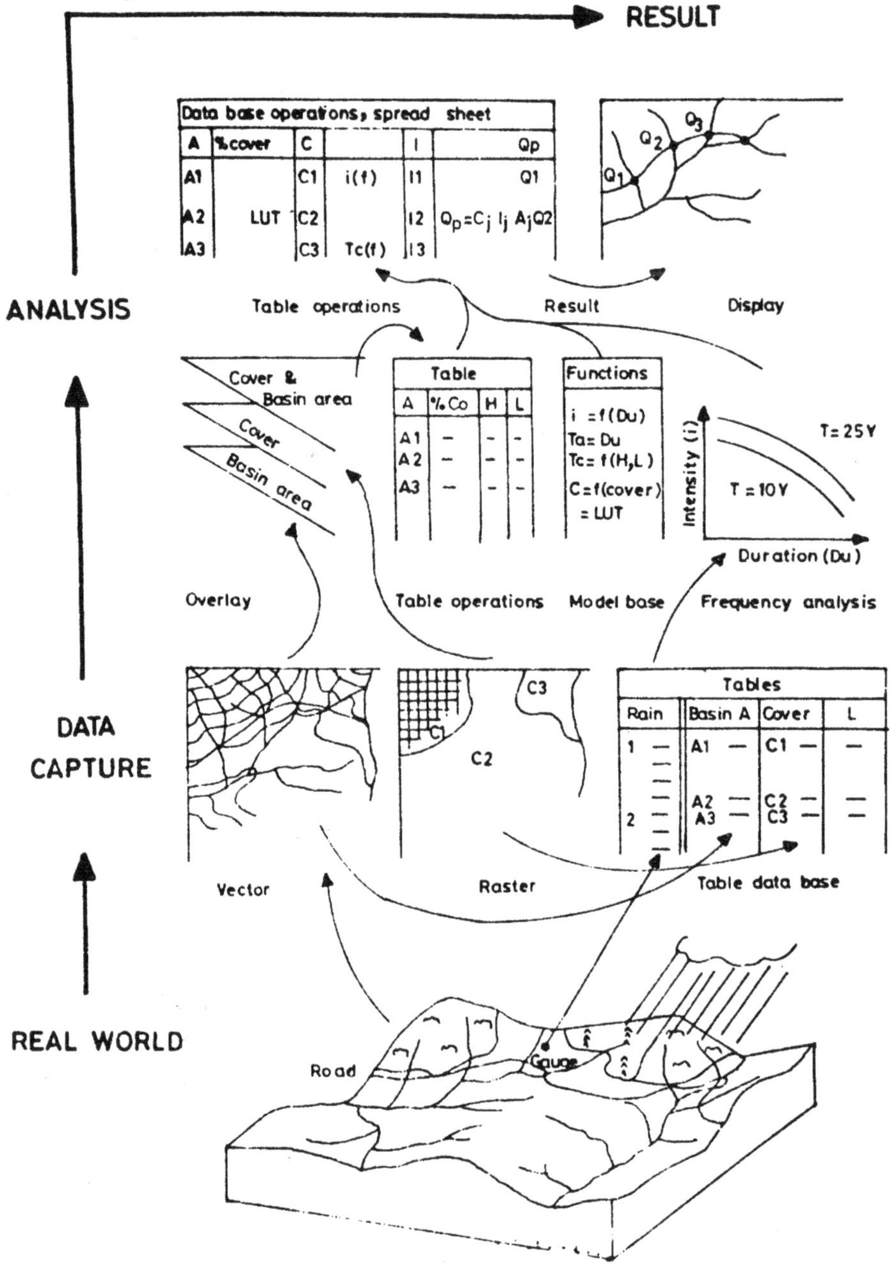

Fig. 12.8. Data flow in GIS from real world situation to desired product

For estimating soil loss, Universal soil loss equation (Eq. 7.4) can be used. The parameter C and to a lesser extent P can be evaluated based on land use/cover information derived from RS. K factor depends on physiography-cum-soil map of the watershed which can be developed using RS and GIS. The R factor can be evaluated using equation 7.5 while SL factor can also be worked out using equation 7.7. C factor represents the ratio of soil loss under a given crop to that from a bare soil. Since crops vary with the season, it should be worked for each season at each crop growing stage viz; seeding, establishment, growing and maturity and after harvest.

The watershed can be prioritised based on soil loss as under:

Soil loss	*Category*	*Priority class*
> 45	Very high	I
35-45	High	II
25-35	Medium	III
15-25	Low	IV
< 15	Very low	V

Ground Water Targeting

Preliminary investigation on ground water exploration using remote sensing helps in delineating potential areas of water bearing strata where exploration work could be started. Using image interpretation keys such as tone, texture, pattern, drainage, topography, etc; satellite images can be used to delineate geomorphological features like palaceo-channels, older flood plains, valley fills, alluvial plains, pediments, lineaments, weathered zones etc. Satellite imageries have been particularly useful in identifying linear features such as fractures, faults, etc; which are ideal localisers of ground water sinks especially in the hard rock terrains. Lithologic contacts are also brought out fairly well in the images. The information on the drainage pattern and drainage density provided by remote sensing have a bearing on the recharge condition and permeability of the rocks. Dykes which act as ground water barrier can be easily identified on the imagery.

Land Capability Classification

Land capability classification is very important for optimal and sustainable utilization of resources. It is an expression of the effect of physical land conditions including climate on the total suitability for use without damage for crops, grazing, fuel and wildlife. Land irrigability classes is made based on the quality and quantity of irrigation water, drainage requirement, topography, crops, soil and economic conditions. Remotely sensed data when used in combination with the conventional data can be an effective tool to arrive at the capability classes (chapter 3).

The data acquisition for watershed treatments is done following two tyre system. At first level, rapid reconnaissance survey is done in the catchment area to identify and demarcate the vulnerable watersheds on the basis of the sediment yield or runoff generation potential. The reconnaissance survey is done in 1:50000 scale using Survey of India topographic maps. The steps followed are:

(i) Generation of drainage map on 1:50000 scale using Survey of India maps,

(ii) Delineation and classification upto sub watershed level following stream hierarchy,

(iii) Rapid reconnaissance survey of the area to formulate erosion intensity mapping unit legend followed by mapping,

(iv) Generation of map showing the spatial distribution of erosion intensity unit (EIMU),

(v) Assignment of weightage and delivery ratio to each erosion intensity map.

(vi) Computation of area of each mapping unit and sub-watershed planimetrical and calculation of sediment yield by using the empirical formula:

$$SYI = \frac{(Aei \times Wei \times DRi)}{AW} \times 100 \text{ ———— } 12.2$$

where: SYI = sediment yield index,
Aei = area of erosion intensity mapping,
Wei = weightage value of the i th mapping unit,
DRi = delivery ratio of the i th mapping unit,
AW = total area of sub watershed, and
n = number of mapping units.

(vii) Characterization of sub-watersheds into very high, high, medium, low and very low priority categories with respect to sediment indices.

The detailed soil survey of the priority watershed is done using 1:15000 scale aerial photographs or cadestral map as base material. The level of mapping is phases of a soil series where different phases viz; effective soil depth, surface texture, stoniness, and rockiness and erosion are depicted on the map, in respect of each soil series. The procedure adopted in conducting detailed soil survey using aerial photo interpretation techniques are:

(i) prefield interpretation of aerial photograph,

(ii) collection of ground truth, and

(iii) past field interpretation of actual photographs.

It involves morphological study of the soil profile at representative sites, through traversing of the sample area depending on heterogeneity of soils, extrapolation of the information to the non-sample area and random checking of boundaries.

The data so collected out of rapid reconnaissance survey provide spatial distribution of various soil and land attributes which constitute the erosion intensity mapping units (EIMU). the broad level information on soil, land use/land cover, slope, erosion, physiography, etc; could be deduced from priority delineation map for regional planning.

Appendices

Appendix 1

Water availability period at some locations in Haryana

Standard	*PET*	*Bhiwani*		*Hisar*		*Mahendergarh*		*Sirsa*	
week number	*mm*	*Rainfall & probability*	*Aridity class*	*Rainfall & probability*	*Aridity class*	*Rainfall and probability*	*Aridity class*	*Rainfall and probability*	*Aridity class*
26	42.6	16.0/.43	D	9.8/.35	D	26.8/.62	M	7.1/.50	D
27	39.7	33.4/.63	M	21.5/.69	M	19.7/.57	M	9.8/.40	D
28	39.2	24.1/.53	M	24.9/.64	M	34.7/.48	M	19.6/.45	D
29	38.5	21.8/.67	M	25.2/.60	M	18.7/.52	MD	20.0/.50	M
30	37.0	32.9/.67	M	29.7/.71	M	30.6/.52	M	18.2/.55	MD
31	42.8	17.3/.60	MD	19.8/.66	MD	36.8/.52	M	23.6/.50	M
32	32.1	22.7/.53	M	27.2/.74	M	39.1/.76	H	19.6/.50	M
33	33.9	18.2/.63	M	28.3/.61	M	38.7/.71	H	28.5/.40	D
34	32.1	16.7/.37	D	22.2/.64	M	32.6/.67	H	15.2/.35	D
35	36.8	17.3/.43	D	21.4/.51	M	30.3/.62	M	6.6/.30	D
36	34.5	18.1/.33	D	16.2/.41	D	25.3/.62	M	9.7/.20	D
37	33.2	24.0/.47	D	17.5/.44	D	22.1/.62	M	18.3/.35	D
Total	—		7		9		12		4

Rainfall in mm, D = dry, MD = moderately dry, M = moist, H = humid

Appendix 2

A Soil Profile

A soil is a three dimensional body on the surface of the earth's crust consisting of solid, liquid and gaseous phases, which under favourable conditions supports and nourishes plants. A vertical section of the soil from soil surface to the parent material is called a soil profile (Fig. i). A layer of soil which is almost parallel to the soil surface but differs from adjacent genetically related layers in physical, chemical and biological properties is called the soil horizon. The surface layer of thickness 0-25 cm is called A-horizon below this of thickness 25-75 is called B-horizon or sub-soil. Directly below B-horizon is C-horizon which contains weathered rocks. In a young soil (a zonal soil), the B-horizon is absent but in a mature (zonal) soil, A—and B-horizons are prominent while C-horizon may be absent. Intrazonal soil (saline and alkali soils, hydrophobic soils) develops under

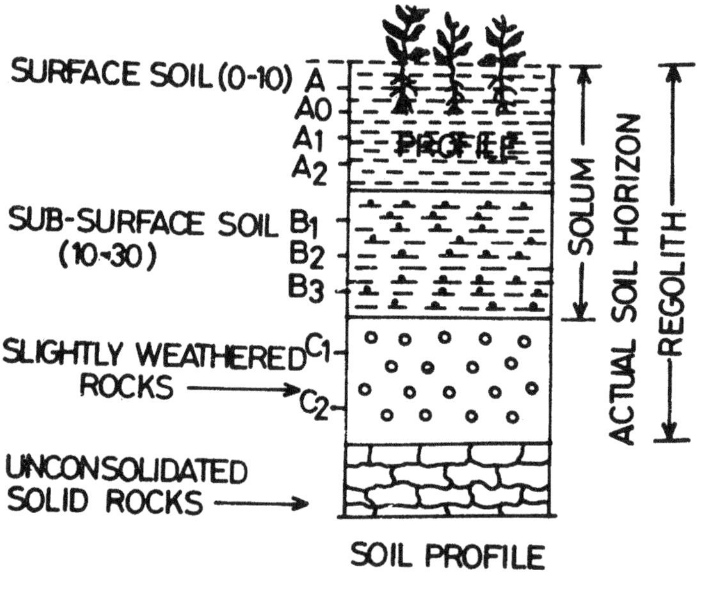

Fig. i. Anatomy of a soil profile

pronounced influence of parent material and local conditions of relief (land slope, water logging etc.) with limited influence of climate and vegetation. The zonal soils are formed under the pronounced influence of climate and vegetation with little effect of relief measures.

Soils could he classified as order, sub-order, great soil group, sub-group (higher order grouping), family, series, types and phase (lower order grouping). A soil series represents a group of soils having soil horizons with similar characteristics and arrangements in a soil profile developed from a particular type of parent material but they differ in the texture of the surface soil. Soil type is the sub-division of soil series based on the variation in the texture of the soil.

Soil Texture

A soil has different sizes of particles. Based on their size they are called sand (2.0 - 0.02 mm), silt (0.02 - 0.002 mm) and clay (0.002 mm). Relative proportion of different sizes of particles in a soil is called soil texture. A soil containing more than 85 per cent sand is called sand, the one containing 40 per cent clay is called clay soil. A loam is a soil which contains almost equal proportion of sand, silt and clay. It is called sandy loam or silty loam or clay loam on the basis of more sand, silt and clay content, respectively. A soil is called, organic if (i) the mineral fraction (sand, silt and clay) has 50 per cent or more clay and there is 30 per cent or more organic matter (ii) the mineral fraction has no clay and there is 20 per cent or more organic matter and (iii) the clay percentage is between 0 and 50 and the organic matter is between 20-30 per

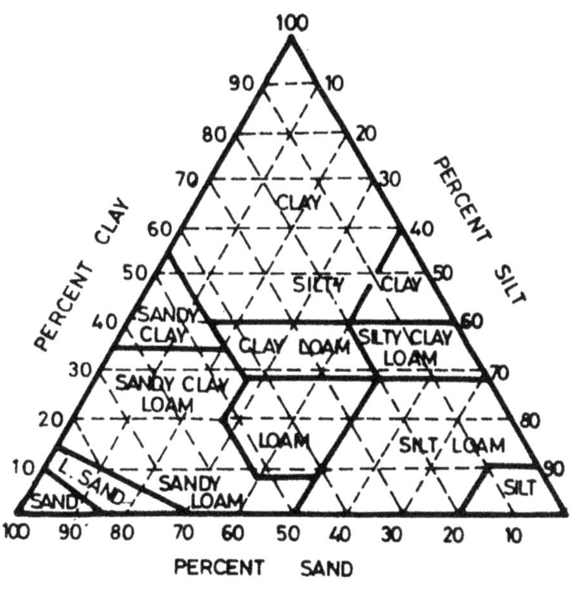

Fig. ii. The soil textural triangle

cent. Sand particles feel gritty to the fingers, silt has the appearance and feel of flour while clay makes soils swell and becomes sticky when wet. The U.S. Department of Agriculture has developed a system of texture designation using the textural triangle (Fig. ii). In using the figure, the points corresponding to the percentage of silt and clay present in the soil under consideration are located on the respective axis. Lines are then projected inward, parallel to clay in the case of silt and parallel to sand in the case of clay. The name of the compartment in which these two lines intersect denotes the textural class of the soil. For example a soil containing 69 per cent sand, 10 per cent silt and 21 per cent clay comes under sandy clay loam texture.

Soil Structure

In a soil the sand, silt and clay particles may bind together by certain binding agents like organic matter, calcium etc. Such bound particles are called soil aggregates or secondary particles. The arrangement of sand, silt and clay with the secondary particles is called soil structure. The structure may be cubical (cube like in shape), columnar (like column), platy (like plate) and spherical (like sphere). The spherical structure may be granular or cumb like. The later is considered desirable for agriculture as it has more pore space.

Water Holding Capacity

Water is held in the soil by surface tension and adsorption forces. Presence of salts in the soil also hold water. The tenacity with which the water is held in soil is called soil water suction and is measured in term of bar. A soil is termed at saturation when all its pores are filled with water and there is no tension on the soil-water. The amount of water held by a soil at saturation is called water holding capacity of the soil. After heavy irrigation to a soil, the soil water drains rapidly for the first few hours and then starts to drain slowly due to gravity. One or two days after irrigation the drainage rate is so slow that in a short course of time it is almost zero. The moisture content of a soil at this level is called field capacity of the soil and it corresponds to about 0.1, 0.3 and 0.5 bar suction respectively for sandy, loam and clayey soils. This is also called the upper limit of plant available soil water. A plant growing at field capacity soil moisture will deplete soil water and the water content of the root-zone will be depleting with time. After a certain period, the soil water content will come down to such a low level that it cannot supply water to plant to meet its needs and the plant starts wilting. The moisture content of the soil at this point is called wilting point moisture. The difference between field capacity and wilting point moisture is called plant available water. For different texture soils, wilting point, field capacity, available water and saturation moisture are illustrated in Table 2.1.

Table 2.1. Field capacity, wilting point and plant available water of different soils

	% volume		
Soil texture	*Field capacity*	*Wilting point*	*Available water*
Loamy sand	18.0	5.5	12.5
Sandy loam	23.0	6.2	16.8
Loam	27.0	7.5	19.5
Silty loam	30.0	11.6	18.4
Clay loam	36.0	19.0	17.0
Clay	40.0	23.5	16.5

The allowable water depletion varies with crop and its variety, type of soil and climatic condition. For wheat in northern India, the allowable water depletion is 30 per cent of the plant available water. In general, irrigation should be applied before half the available water in the root zone is depleted. For irrigation planning this concept is commonly used.

Appendix 3

Proforma for Socio-Economic Survey

Village:———————— Block________ Tehsil_________ District______

1. Particulars of Population

(a) Number of houses..................

(b) Total population.......................Male .... Female...................

(c) Schedule caste.........................Male .... Female...................

(d) Schedule tribes.......................Male .... Female...................

(e) Backward class........................Male .... Female...................

(f) Farming families......................Self farming......... Landless labour......

(g) Literacy..................................Male .... Female...................

(h) Enterprises (i) Farming..........% (ii) Service..........% (iii) Other...........%

(i) Annual income per person................

(j) Major expenses: (i) Food........ (ii) Clothing........... (iii) Housing..........
(iv) Education.............. (v) Health............. (vi) Social Function.........
(vii) Entertainment........... (viii) Others.............

2. Institutions

Name of Institute	Name/Location
(a) Village Panchayat	....
(b) School	....
(c) Cooperative Society	....
(d) Hospital	....
(e) Veterinary Hospital	....
(f) Bank	....
(g) Agriculture Department	....
(h) Agricultural Research	....
(i) ICDP (Integrated Child Development Programme)	....
(j) Voluntary Organization	....

3. Land

(a) Total land area........................ ha.

(b) Land under cultivation........... ha.

(c) Wasteland................................ ha.

(d) Culturable wasteland............ ha.

(e) Land under (i) Forest.......... ha (ii) Horticulture................. ha (iii) Pasture........... ha.

(f) Irrigated land.......... ha (i) Tubewells................ % (ii) Canals.......... %

(g) Soil texture (i) Light................ (ii) Medium............ (iii) Heavy ....

(h) Soil Fertility: Nitrogen.......... Phosphorus.......... Potash......... kg ha^{-1}

(i) Major crops and their average yield (q ha^{-1})

Summer crop *Yield* *Winter crop* *Yield*

(j) Animals (i) Milching................ (ii) Draft................. (iii) Others............ Cow..................... Buffalo....................

(k) Farm machineries (i) Tractor.......... (ii) Thresher.......... (iii) Others........

4. Developmental Activities

Institution Activity Coverage Year of Start

5. Climatic:

(i) Annual rainfall.......... mm (ii) Seasonal rainfall: Summer....... mm, winter.......... mm (iii) Rainy season duration......... weeks (iv) Peak rainfall and month.................. mm (v) Peak rainfall intensity.............. mmh^{-1} (vi) Potential evapo-transpiration............. mm (vii) Annual and seasonal open pan evaporation .... mm, (viii) Expected runoff from a watershed........... mm, total................. m^3, Peak................. m^3/second.

6. Major Production Problems and Redressing Activities

(a) Agriculture Production

(i) (iv)

(ii) (v)

(iii) (vi)

(b) Veterinary

(i)

(ii)

(c) Human Resource Development

(i)

(ii)

(d) Natural Resource Management

Appendix 4

Master Plan of Bunga Watershed

Introduction

The Bunga Watershed is typical of the Shiwalik foothill region having problems of over grazing and large-scale denudation. People are poor and mostly depend on hills for their livelihood. The rains are intense, excess and runoff producing in summer while the winter season receives very low rainfall. Dryspells are frequent and the land is slopy or undulating. Drinking water is also available with difficulty. Rearing of cattle is the most important profession. For socio-economic development of people soil and water conservation measures coupled with modern agriculture is essential. Also livestock farming needs to be improved by improving the breed and reducing the number of cattle.

Location

The project area (463 ha) is located at village Bunga in district Panchkula on Panchkula—Ambala road about 35 km from Chandigarh. The other villages falling in the project are Belwali, Dabkodi, Behar, Bellanti and Kot.

Climate

The climate of the area is subtropical with an average annual rainfall of 1,115 mm. Eighty per cent of annual rainfall is received from June to September. The rainfall intensity of 100 mm h^{-1} for 0.25 hour is common. Only 141 mm rainfall is received in the winter season from November to March. The temperature has a wide range of 0 – 45°C. Lowest temperature of 0°C in winter (January) and highest temperature of 45°C in summer could be recorded occasionally.

Soils and Vegetation

The soils are made up of sand stones, grits and clays and are derived from young and loose rocks which are susceptible to erosion. The texture of soils is silty loam in the catchment and loamy sand to sandy loam in the down stream in the agricultural land. The vegetation is sparse with trees like *Acacia catechu*, *Dalbergia sissoo*, *Adhatoda vasica*, *Carissa apaca* and *Dodonaea viscosa* and grasses like *Dichanthium annulatum*, *Eulaliopsis binata* and *Chrysepogon fulvus*. Permanent vegetation is almost absent.

Land Capability Classification

The capability wise distribution of land area is as under:

Class	*Area (ha)*	*Total (%)*
I	36.8	7.8
II	31.1	6.7
III	313.4	67.7
Catchment Area	127.0	

Out of total cultivable land of 381 ha, 243 ha could be provided with irrigation and the remaining 138 ha is left out for dryland agriculture.

Land Holdings and Livestock

There were 422 families out of which 301 families owned land and rest of the 121 families were self-employed (50%) and the remaining were labourers. The people kept large herd of goats and cows which grazed unrestricted in the adjoining hills, leading to diminishing vegetation and land degradation.

Soil and Water Conservation Measures

For arresting soil erosion and rainfall runoff mechanical measures like water storage reservoir, water conveyance system, bunding, staggered contour trenches and stone check dams at appropriate sites were taken up. The storage reservoir had following specifications:

Catchment area	-	127 ha
Bed level at site	-	100 m
Inlet level of pipe outlet	-	104 m
Full supply level	-	113 m
High flood level	-	114 m
Crest elevation	-	116 m
Top width	-	5 m
Side slope	-	3:1
Submergence area	-	10.0 ha
Total storage capacity	-	60.0 ha-m
Live storage capacity	-	57.3 ha-m
Dead storage capacity	-	2.1 ha-m

The key trench of the earthen dam had following specifications

Bottom width	-	3 m
Depth of trench	-	4.5 m
Top width	-	12.0 m
Side slopes	-	1:1

The hilly catchment area of 46.0 ha was treated with gully plugs (1400 small) and gabion structures (4) besides plantation of various trees and grasses. The entire catchment area was protected against grazing and cutting of trees. The villagers were encouraged not to allow grazing but cut grasses for stall feeding their cattle. The staggered contour trenches of 2.5 m × 0.45 m × 0.45 m at a distance of 2 m in the same line and 2.5 m vertical interval were made to control soil erosion and improve rainfall conservation *in-situ* for quick growth of trees.

The stored water in the reservoir was transported to the outlet for irrigating crops grown in 243 ha land by an underground pipeline of diameter 20.0 cm laid at 104 m RL. From the outlet tanks, water was conveyed to fields through PVC pipes of different sizes.

Dryland Agriculture

Maize-Chickpea/*Taramira*, Fallow-Wheat, Groundnut-Chickpea, Pigeonpea-Chickpea crop rotations were followed by the farmers using improved agronomic practices. Forest trees like *Eucalyptus* hybrid, *Dalbergia sissoo* and *Acacia catechu* and grasses like *Eulaliopsis binata* and *Terminalia arjuna* were planted on berm of contour trenches and on both sides of *cho* for training flash floods and excess water to a disposable site.

Appendix 5

Suitable species of trees, shrubs and grasses for dry and salt affected lands

Dryland	*Salt affected land*
	Trees
Acacia arabica (Deshi babul)	*Acacia nilotica* (Deshi babul)
Acacia catechu (Khair)	*Acacia tortilis* (Israili babul)
Acacia juliflora (Vilayti babul)	*Albizia lebbek* (Siras)
Acacia senegal (Kumbhat)	*Atriplex nummularia* (Buee)
Acacia tortilis (Israeli babul)	*Azadirachta indica* (Neem)
Ailanthus excelsa (Ardu)	*Dalbergia sissoo* (Sisham)
Albizia lebbek (Sirash)	*Leucaena kucocephala* (Subabul)
Azadirachta indica (Neem)	*Phoenix sylvestris* (Khajur)
Caparis aphylla (Kher)	*Prosopis juliflora* (Vilayati babul)
Cassia siamea (Kassod)	*Salvadora oleoides* (Pilu)
Parkinsonia aculeta (Vilayti kiker)	*Salvadora persica* (Jal)
Populus sp. (Poplar)	*Tamarix articulata* (Pharas)
Prosopis cineraria (Khejri)	
Prospis juliflora (Vilayati babul)	
Prosopis spicigera (Desi khejri)	
Tamarix aphylla (Lal Jhau)	
Tamarix articulata (pharas)	
Tamerindus indica (Imali)	
Tecomella undulata (Roheda)	
	Shrubs
Caparis decidua (Kair)	*Capparis decidua* (Kair)
Sancchrum munja (munj)	*Halaxylon salicornicum* (Lama)
Zyzyphus nummularia (Bordi)	*Sancchrum munja* (Munj)
	Suaeda fruticosa (Lonia)
	Tamarix ericoides (Jhau)
	Zizyphus nummularia (Bordi)
	Grasses
Cenchrus ciliaris (Anjan)	*Chloris gayana* (Rhodes)
Cenchrus setigerus (Dhaman)	*Cynodon dactylon* (Barmuda)
Culalfopsis binata (Bhabhar)	*Diplechue fusca* (Karnal)
Cynodon dactylon (Barmuda)	*Penicum antidotal* (Blue)
Dichanthium annulatum (Dub)	*Penicum leavifolium* (Gunada)
Lesirus sindicus (Sewan)	*Sporbolus diander* (Chiriyaka dana)
Vetiveria zizanoides (khas)	Terminalia arjuna (Arjun)

Appendix 6

Average daily water requirement of animals and human beings

Animal/Human beings	*Water requirement Litre/day*
Buffalo	10-12
Camel	15-20
Cow	10-12
Donkey	8-10
Goat	3-4
Sheep	4-5
Human beings (Rural)	45 (Domestic and drinking uses)

Appendix 7

Discharge of Parshall flume of different widths and flow head differences

Head Difference (cm)	Width, cm									
	7.6	15.2	30.5	50	75	100	125	150	175	200
2	0.4	0.2	–	–	–	–	–	–	–	–
5	1.7	3.3	7.1	11.4	16.7	21.8	26.8	36.5	41.3	50.8
10	5.0	10.0	20.4	33.3	49.1	64.6	80.0	110.2	125.1	154.8
15	9.4	19.0	37.9	62.3	92.3	122.1	151.7	210.2	239.2	297.1
20	14.6	30.0	58.7	97.1	144.5	191.8	238.9	332	379	472
25	20.7	42.6	82.4	136.9	204.6	272.2	340	474	541	676
30	27.4	56.7	108.8	181.4	271.8	362	433	634	725	906
40	–	89.6	168.6	282.6	426	569	714	1003	1148	1438
50	–	–	236.8	399.0	603	808	1013	1431	1640	2053
60	–	–	312.0	528.0	800	1076	1354	1913	2195	2760
70	–	–	–	–	–	1371	1727	2445	2808	3537

Bibliography

Abrol, I. P. and Sehgal, J. L. 1994. Degraded land and their rehabilitation in India. *In: Soil Management for Sustainable Agriculture in Dryland Areas*. Bulletin No. **16**, ISSS, New Delhi.

Bhattacharjee, J. G., Raychowdhury, C., Landey, R. L. and Pandey, S. 1982. *Bioclimatic Analysis of India*, NBSSLUP, Nagpur, Bulletin, **7**.

Bhumbla, D. R. 1994. Soil and water management in India-some key issues. *In: Management of Land and Water Resources for Sustainable Agriculture and Environment*, Bulletin No. **16**, ISSS, New Delhi.

Erasmus, I. J. and Bansal, R. C. 1985. A runoff prediction for use in watershed planning in the Ambala Sivaliks. *Indian Forester*, **91**: 548-52.

Koppen, W. 1931. Outline of Climatology, Water de Gryter Co; Berlin.

Meher Homji, V. M. 1975. In: *Environmental analysis of the Thar Desert* [ed. Gupta, R. K. and Ishwar Prakshl], English Book Depot, Dehradun.

Raman, C. V. R. and Shrinivashan Murthy, B. 1971. Evaluation of crop growing period based on water balance. *Sci. Rep.* **173**, IMD, Pune.

Sehgal, J. L., Mandal, D. K., Mandal, C. and Vadivelu, S. 1992. *Agro-ecological Regions of India.* Oxford and IBH Publishing Co. [P] Ltd; New Delhi.

Singh, R. P. and Subha Reddy, G. 1988. In: *Drought research priorities for the dryland tropics*, [ed. Bidinagar, F. R. and Johnson, C.], ICRISAT, Patancheru, Hyderabad.

Subramaniam, S. and Samuel, J. C. 1990. Soil conservation in river valley projects in India. Proc. International symposium on water erosion, sedimentation and resource conservation, held at CSCRATI, Dehradun.

Tejwani, K. G. 1975. Soil and water conservation research-1956-71, ICAR, New Delhi.

Thornthwaite, C. W. 1948. An approach towards a rational classification of climate. *Geogr. Rev*; **38**: 55-94.

Glossary

Barley	*Hardeum vulgare*
Black gram	*Phaseolus mungo*
Castor	*Richinus cummunis*
Chick pea	*Cicer arietinum*
Cluster bean	*Cyamopsis tetragonoloba*
Cotton	*Gossypium arboreum*
Cowpea	*Vigna sinesis* (L.)
Finger millet	*Eleusine corasana*
Groundnut	*Arachis hypogaea* (L.)
Horsegram	*Macrotyloma ciniflorum*
Kidney bean	*Phaseolus aconitifollus* Jacq
Lentil	*Lensculinaris*
Maize	*Zea-mays*
Mung	*Phaseolus aureus*
Mustard	*Brassica juncea* (L.)
Oat	*Avena sativa*
Pea	*Pisun* sp.
Pearl millet	*Pennisetum typhoides*
Pigeon pea	*Cajanus cajan*
Potato	*Sechiumedule*
Safflower	*Carthamum tinctorius*
Sesamum	*Sesamum indicum*
Sorghum	*Sorghum bicoler*
Soybean	*Glycine max.* (L.)
Sunflower	*Halianthus annuus*
Tabacco	*Nicotiana tabacum*
Wheat	*Triticum estivum*
Taramira	*Eruca sativa*

Index

www.ingramcontent.com/pod-product-compliance
Ingram Content Group UK Ltd.
Pitfield, Milton Keynes, MK11 3LW, UK
UKHW060119010826
14089UKWH00005B/69

9 789354 615436